This authoritative volume provides a comprehensive review of the origin and evolution of planetary nebulae. It covers all the stages of their evolution, carefully synthesizes observations from across the spectrum, and clearly explains all the key physical processes at work. Particular emphasis is placed on recent observations from space, using the Hubble Space Telescope, the Infrared Space Observatory, and the ROSAT satellite. This book presents a thoroughly modern understanding of planetary nebulae, integrating new developments in stellar physics with the dynamics of nebular evolution. It also describes exciting possibilities such as the use of planetary nebulae in determining the cosmic distance scale, the distribution of dark matter, and the chemical evolution of galaxies.

This book provides graduate students with an accessible introduction to planetary nebulae, and researchers with an authoritative reference. It can also be used as an advanced text on the physics of the interstellar medium.

T0211412

THE ORIGIN AND EVOLUTION OF PLANETARY NEBULAE

Cambridge astrophysics series

Series editors

Andrew King, Douglas Lin, Stephen Maran, Jim Pringle and Martin Ward

THE ORIGIN AND EVOLUTION
OF PLANETARY NEBULAE

SUN KWOK

University of Calgary, Canada

CAMBRIDGE
UNIVERSITY PRESS

CAMBRIDGE UNIVERSITY PRESS
Cambridge, New York, Melbourne, Madrid, Cape Town, Singapore, São Paulo

Cambridge University Press
The Edinburgh Building, Cambridge CB2 8RU, UK

Published in the United States of America by Cambridge University Press, New York

www.cambridge.org
Information on this title: www.cambridge.org/9780521623131

First published 2000
This digitally printed version 2007

A catalogue record for this publication is available from the British Library

Library of Congress Cataloguing in Publication data
Kwok, S. (Sun)
The origin and evolution of planetary nebulae / Sun Kwok.
      p.   cm. – (Cambridge astrophysics series : 33)
ISBN 0-521-62313-8 (hc.)
1. Planetary nebulae.   I. Title.   II. Series.
QB855.5.K96   1999
523.1′135 – dc21   99-21392
                                                         CIP

ISBN 978-0-521-62313-1 hardback
ISBN 978-0-521-03907-9 paperback

Dedicated to my father: Chuen Poon Kwok

Contents

Preface

This book reflects the extraordinary amount of progress made in planetary nebulae research in the last thirty years. Before 1970, observations of planetary nebulae were limited to the visible region, and theoretical understanding focused on the physical processes in the ionized region. As the result of observations across the electromagnetic spectrum, we now have a much better appreciation of the richness of the planetary nebulae phenomenon. All states of matter (ionized, atomic, molecular, and solid state) are present in planetary nebulae, emitting radiation via a variety of mechanisms. More importantly, we have achieved a much better understanding of the origin and evolution of planetary nebulae (hence the title of the book).

When I was first approached by the Cambridge University Press about the possibility of writing a book on planetary nebulae, I as initially hesitant given the heavy teaching and administrative duties that I have at the University. In the end, I am glad to have done it because it offered me relief from writing reports and doing budgets as well as the opportunity to organize my own thoughts on the subject. The task of writing was made easier because of the availability of software tools: the manuscript was written in CUP LATEX, the calculations performed using MATHCAD, and many of the figures prepared using AXUM.

Most of the writing was done during my sabbatical year spent at the Institute of Astronomy and Astrophysics of Academia Sinica (Taiwan) and the Smithsonian Astrophysical Observatory; I thank Fred Lo, Typhoon Lee, Paul Ho, and James Moran for their hospitality. I would also like to thank Orla Aaquist, Peter Bernath, Bill Latter, Detlef Schönberner, Ryszard Szczerba, and Albert Zijlstra who read earlier drafts of the manuscript and gave me valuable comments. Several of the beautiful planetary nebulae images included in the book came from Trung Hua, who also contributed to the instrumentaion section. Chris Purton gave the manuscript a thorough reading and pointed out numerous errors that had escaped my notice. My colleagues Kevin Volk, Tatsuhiko Hasegawa and Cheng-Yue Zhang helped with various sections of the book. I am particularly indebted to Professor Lawrence Aller not only for his detailed feedbacks on the manuscript, but also for his books from which I learned much of this subject. I want to thank my family for their support and in particular my daughter Roberta who helped prepare and check the references.

S.K.
Calgary, Canada
May 1999

1

History and overview

The first planetary nebula was observed by Charles Messier in 1764 and was given the number 27 in his catalog of nebulous objects. The final version of the Messier catalog of 1784 included four planetary nebulae (PN) together with other nonstarlike objects such as galaxies and star clusters. The name planetary nebulae was given by William Herschel, who found that their appearances resembled the greenish disk of a planet. With better telescope resolution, nebulae that are made up of stars (e.g., galaxies) were separated from those made up of gaseous material. PN were further distinguished from other galactic diffuse nebulae by that fact that PN have definite structures and are often associated with a central star. This distinction became even clearer with spectroscopy. The first spectrum of a PN (NGC 6543) was taken by William Huggins on August 29, 1864. The spectra of PN are dominated by emission lines, and not a continuous spectrum as in the case of stars. The first emission line identified was a Balmer line of hydrogen (Hβ), although stronger unidentified lines could be seen in the spectrum. Since the spectra of PN are entirely different from those of stars, their luminosity cannot be due to reflected starlight.

The idea that PN derive their energy from a nearby star was first considered by Herschel (1791). However, no further progress was made for another century. Hubble (1922), using data obtained with the Mount Wilson 60- and 100-in. telescopes, found a correlation between the magnitude of the central star and the size of the nebula. He therefore argued that the emission-line spectrum of PN is the result of the nebula absorbing the continuous radiation from the central star. In order to explain the strength of the Hβ line, Menzel (1926) suggested that all the stellar output beyond the Lyman limit (912 Å) must be utilized to ionize the hydrogen (H) atom. The mechanism that the lines of hydrogen and helium (He) are emitted as the result of recombination between the nucleus and electron after the nebula is photoionized was quantitatively developed by Zanstra (1927). Most importantly, Zanstra was able to determine the number of Lyman continuum photons emitted from the observed ratio of the Balmer line to stellar continuum flux, and was therefore able to deduce the temperature of the central star (see Section 7.1.1). The central stars of PN were found to have very high temperatures, which were much hotter than those of any other known stars at the time.

However, a number of strong nebular lines remained unidentified by laboratory spectroscopy and were suggested to be due to some unknown element "nebulium." The strength of the lines led to the conclusion that these lines must originate from known elements of high abundance but are emitted under unusual conditions. One such condition is the low density of the interstellar medium. Russell *et al.* (1927) speculated that certain atoms with metastable states, which do not have the time to emit radiation because of collisional deexcitation in the high-density terrestrial environment, will radiate under interstellar conditions. Bowen, in 1928, identified eight of the strongest nebular lines as being due to metastable states of N^+ (singly ionized nitrogen), O^+, and O^{++}. These metastable states lie a few electron volts above the ground state and can be collisionally excited by electrons freed by the photoionization of hydrogen. The presence of highly excited, strong optical lines of oxygen was explained by Bowen (1935) as being the result of a fluorescence mechanism.

Since the forbidden lines are collisionally excited, and therefore remove energy from the kinetic energy pool of the electrons, they represent a major source of cooling of the nebula. Menzel and Aller (1941) were able to show that, no matter how hot the central star, cooling by the forbidden lines limits the electron temperature to <20,000 K.

Observations with better spectral resolution led to the discovery that the emission lines in PN are broad, or even split. This was correctly interpreted as expansion, and not rotation, of the nebula (Perrine, 1929). With the adoption of a size of 0.3 pc and an expansion velocity of 30 km s^{-1}, the dynamical lifetime of the PN can be estimated to be $\sim 10^4$ yr.

1.1   Planetary nebulae as a phase of stellar evolution

At the beginning of the 20th Century, when stars were believed to evolve from high temperatures to low temperatures, PN were thought to be very young stars because of their high temperatures. From his studies of the velocity distribution of PN, Curtis (1918) found that PN are more similar to late-type stars and are unlikely to be young objects. Theoretical understanding of the origin of PN began with Shklovsky (1956b), who suggested that PN are progenitors of white dwarfs (WDs) and descendants of red giants. By tying PN to red giants and white dwarfs, Shklovsky recognized that these stars must be evolving rapidly. This view was supported by Abell and Goldreich (1966) who used the expansion velocities of PN and the escape velocities of red giants to argue that PN are the ejected atmospheres of red giants. Using the total number of galactic PN of 6×10^4 (as estimated by Shklovsky) and a lifetime of 2×10^4 yr, Abell and Goldreich showed that PN must be forming at a rate of 3 per year. Since this is of the same order as the number of stars leaving the main sequence, they suggested that practically all low-mass stars will go through the PN stage. This established the importance of PN in the scheme of stellar evolution.

Although Shklovsky successfully drafted a qualitative scenario for PN evolution, the details of the transition from red giants to PN to WDs remained very poorly known for another 20 years. For example, in the 1960s it was commonly believed that the horizontal branch was an essential phase of the evolution of low mass stars. However, the way that PN are related to horizontal branch stars was not at all clear.

In this book, we present a modern view of the origin and evolution of PN, tracing their origins to the mass loss on the asymptotic giant branch (AGB). The circumstellar envelopes that are created by the mass-loss process over a period of 10^6 yr are swept up by a new fast stellar wind into the shell-like structure that we observe in PN. The

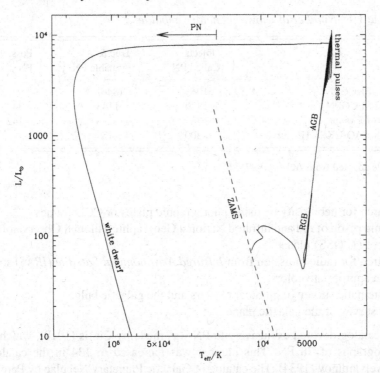

Fig. 1.1. The evolutionary track of a $3M_\odot$ star on the H-R diagram beginning from the zero age main sequence (ZAMS), through the red giant branch (RGB) and AGB to PN and ending as a white dwarf (Figure from T. Blöcker).

interaction of the two winds creates a high temperature-bubble that exerts pressure on the shell and causes it to expand. The core of the AGB star, having lost its envelope, evolves to the higher temperature as its thin remaining H envelope is burnt up by nuclear processes. The increasing output of UV photons will gradually ionize the shell, which at the same time is growing in mass as more AGB wind material is swept up. When the stellar H envelope is used up by nuclear burning, the core will cool down and decrease in luminosity to become a white dwarf (Fig. 1.1).

1.2   Discovery and identification

Based on their diffuse appearances, PN were first cataloged together with galaxies and clusters as part of the New General catalog of Clusters and Nebulae (NGC) in 1887. Many PN carry their NGC designations to this day. In the 20th Century, new PN were discovered either by their appearances on photographic plates or their emission-line spectrum. For example, PN were identified by Abell, using photographs obtained with the Palomar 48-in. Schmidt telescope, and by Minkowski, using objective prism plates taken with the Mt. Wilson 10-in. telescope. Examinations of the Palomar Atlas by Abell (1966), Kohoutek, and others have led to the identification of large numbers of PN. Through objective prism surveys, hundreds more PN were found by Minkowski (1964), Henize (1967), and Thé (1962). More recent discoveries of PN have used a number of methods:

Table 1.1. *Number of known planetary nebulae*

Year	Objects Called PN	True & Probable PN	Possible PN
1967 PKCGPN	1063	(846)	
1992 SECGPN	1820	1143	347
1996 *1st suppl.*	+385	+243	+142
1998 AAO/UKST Hα survey	+>300	+>150	

Table adapted from Acker (1997).

- Search for nebulosity by using photographic plates or CCD frames
- Comparison of red and infrared National Geographic-Palomar Observatory Sky Survey (POSS) plates
- Search for radio emission from *Infrared Astronomical Satellite (IRAS)* sources with appropriate colors
- Systematic survey of globular clusters and the galactic bulge
- Hα survey of the galactic plane

The first catalog devoted exclusively to PN was made by Curtis (1918), which contained photographs of 78 PN. This number was increased to 134 in the catalog of Vorontsov-Velyaminov (1934). The catalog of Galactic Planetary Nebulae by Perek and Kohoutek in 1967 has over 1000 PN included. In the Strasbourg-ESO PN catalog (Acker *et al.*, 1992), 1,143 objects are listed as true PN, 347 as possible PN, and another 330 as mis-classified PN. Since the publication of the Strasbourg-ESO Catalog, a number of PN candidates, selected based on their *IRAS* colors, have been confirmed by optical and radio observations (van der Steene *et al.*, 1995, 1996). These and other new PN are included in the First Supplement to the Strasbourg-ESO Catalog of Galactic PN (Acker *et al.*, 1996). In the Anglo-Australian Observatory/UK Schmidt Telescope Hα survey of the southern galactic plane, many new faint and extended PN were discovered (Parker and Phillipps, 1998). When this survey is completed, there is a potential of nearly doubling the number of PN cataloged. A summary of the numbers of known PN is given in Table 1.1 and a plot of the galactic distribution of PN is shown in Fig. 1.2.

The numbers in Table 1.1, however, do not represent the total population of PN in the Galaxy. Many PN are hidden by interstellar extinction in the galactic plane, and most of the PN on the other side of the galactic center are not seen. Old PN have a very low surface brightness and are difficult to identify. Distant PN are stellar in appearance and cannot be easily distinguished from stars. It is estimated that the total number of PN in the Galaxy can be 10 times higher (see Chapter 18).

1.3   Confusion with other galactic objects

The identification of PN is based on a combination of morphology (shell plus central star) and spectroscopy (strong emission line spectrum with little or no continuum, see Fig. 1.3). The most common confusing sources are emission-line galaxies, reflection nebulae, HII regions, symbiotic stars, M stars, and other emission-line stars. For example,

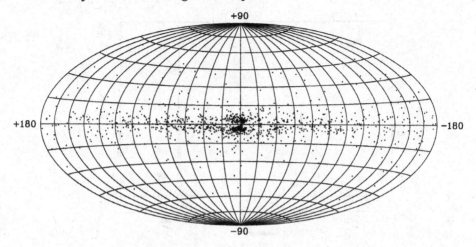

Fig. 1.2. The galactic distribution of PN in the Strasbourg-ESO catalog of Galactic PN.

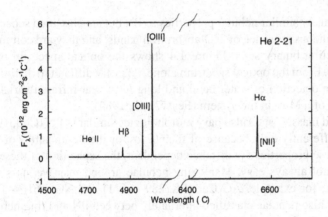

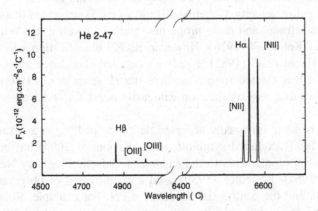

Fig. 1.3. Top: typical optical spectrum of PN. Bottom: optical spectrum of a low-excitation PN (data from T. Hua).

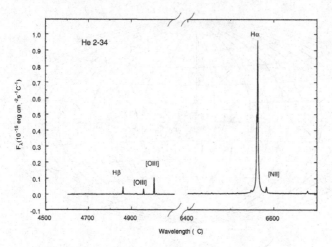

Fig. 1.4. Optical spectrum of a symbiotic star.

symbiotic stars show many similar nebular properties as PN but are classified separately because of the simultaneous presence of TiO absorption bands, and they are commonly believed to be interacting binary stars. Figure 1.4 shows the optical spectrum of the symbiotic star He2-34. From the optical spectrum alone, it is very difficult to distinguish it from a PN. Only the detection of water band and long-term near-infrared variability confirms the presence of a Mira in the system (Feast *et al.*, 1983).

Ring nebulae around massive stars also have morphologies similar to PN (Chu, 1993). They are classified differently only because of their sizes, or the luminosities of their central stars. Since these properties are distance dependent, the separation of these two classes of objects is not always easy. Many ring nebulae around massive stars were once cataloged as PN, for example, AG Car (PK 289 − 0°1) and NGC 6164-5 (PK 336 − 0°1). There are also nebulae classified alternately between PN and ring nebulae, e.g., M1-67 and We21. In the case of M1-67, it was first classified as a HII region by Sharpless (1959) but was included in the PN catalog of Perek and Kohoutek (1967) based on its high heliocentric velocity. Cohen and Barlow (1975) suggested that it is a ring nebula based on its infrared and radio properties, and as a result it was removed from the PN catalog by Kohoutek (1978). However, its PN classification was again suggested by van der Hucht *et al.* (1985). Finally, a detailed abundance analysis of the nebula confirms that it is ejected from a massive star (Esteban *et al.*, 1993). The example of M1-67 shows that misclassification can easily occur for less well-studied objects.

Unfortunately, there is not a universally accepted definition of PN. As an example, whereas Acker *et al.* (1992) excluded symbiotic stars, Kohoutek (1994) has continued to include them in his supplements to the catalog of Galactic Planetary Nebulae (Perek and Kohoutek, 1967). Kohoutek (1989) used a combination of observational properties of the nebula and the central star to define a PN. For example, Kohoutek places density, size, and expansion velocity ranges on the nebula, and temperature, luminosity, and gravity limits on the central star in order for an object to qualify as PN. These observational definitions not only reflect properties commonly observed in

PN, but also represent the imposition of our theoretical understanding of the phenomenon. Clearly observational criteria alone are not sufficient and a combined approach is necessary.

One could go a step further and define PN as ionized circumstellar shells showing some degree of symmetry surrounding a hot, compact star evolving between the AGB and WD phases. Even in such a restrictive definition, it is still not clear how binary stars fit in. For example, PN with binary nuclei can go through mass transfer followed by thermonuclear ignition, which makes them very similar to symbiotic stars or novae. One or more mass transfer phases can occur in an interacting binary system, leading to many different evolutionary scenarios. Mass loss can occur under some of these scenarios, resulting in a PN-like object. For example, the nebular spectra of the symbiotic stars V1016 Cygni and HM Sge are believed to be due to the ionization of the Mira stellar wind by the companion white dwarf. In this book, I will avoid these complications and concentrate the discussion on single star evolution.

1.4   Plantary nebulae as a physics laboratory

PN present an ideal laboratory for the study of the interaction between radiation and matter. The system is simple. All the energy of a nebula is derived from a single source, the central star. Radiation emitted by the star is absorbed and processed by the nebula, which contains matter in ionized, atomic, molecular, and solid-state forms. Because early (pre-1970s) observations of PN were limited to the visible region, our knowledge was restricted to the ionized gas component. Through active interactions between atomic physics and nebular observations, considerable progress has been made. For example, nebular densities and temperatures can be measured by comparing the strengths of forbidden lines (see Section 3.5). However, such determinations depend on accurate values for the spontaneous decay rates and the collisional cross sections. Since forbidden lines include magnetic dipole and electric quadrupole transitions, the observations of these nebular lines stimulated the calculations of the wave functions of multielectron atoms and ions and the corresponding transition probabilities (Shortley *et al.*, 1941; Aller *et al.*, 1949). Applying the techniques of quantum mechanical scattering theory, Seaton (1954b) calculated the collisional cross sections of many ions.

With the use of the assumption of Zanstra, that all the Lyman photons are absorbed in the nebula, the relative strengths of the Balmer lines can be determined by solving the equation of statistical equilibrium if the spontaneous decay and recombination rates are known. The early work of Plaskett (1928) contained only seven levels. This was later improved by Menzel and Baker (1937), who set up an exact algebraic solution to the equations. Consequently, the relative intensities of the Balmer lines can be calculated by approximation techniques and can be compared with observations. With greatly improved computing capabilities in the 1960s, the theory of Balmer decrement was developed to a high degree of accuracy (Brocklehurst, 1970).

The confrontation between theory and observations continues as the quality of both nebular spectroscopy and computational methods improve. Early photographic spectrophotometric measurements were improved by photoelectric calibrations. This was followed by the use of the electronic camera, the image tube, the image-tube scanner, and more recently, the charged-coupled device (CCD). Since the early spectroscopic observations of PN by Wright (1918), extensive databanks on emission lines were built up by Aller *et al.* (1955; 1963) and by Kaler *et al.* (1976). Recent advances in CCD technology have made possible high-resolution spectroscopy with high accuracy, and a substantial increase in the quantity of spectral information has been generated. For example, the number of detected and identified lines from the PN NGC 7027 has increased from the ~250 tabulated in Aller (1956) to more than 1,000 (Péquignot, 1997). This advance has created the need for determinations of the energy levels of many new atomic of ions of common elements and their transition rates. The analysis of nebular spectra is now performed by computer codes known as photoionization models. These models store a large amount of atomic data in the code and use certain elemental abundance and stellar and nebular parameters as inputs. The calculated emergent line spectrum is then used for comparison with observations. The continued refinements of the atomic data have led to a reasonable agreement between the photoionization models and the optical spectra of PN.

This book is roughly organized into three parts. In Chapters 2-6 we describe the physics of the nebula. The physical processes in the ionized component are discussed in Chapters 2-4. The physics of neutral gas and dust components, which were discovered as the result of millimeter-wave and infrared observations, is treated in Chapters 5 and 6. The properties of the central stars of PN are summarized in Chapter 7 and the morphologies of PN are described in Chapter 8.

The PN phenomenon has its origin in the preceding stellar evolutionary phase, the AGB. The structure of AGB stars, and in particular the mass loss that occurs in that phase, is described in Chapter 10. The theory of evolution of central stars of PN is summarized in Chapter 11. The effects of AGB mass loss on PN formation and the subsequent dynamical evolution are discussed in Chapters 12 and 13.

The immediate progenitors and descendants of PN, the proto-PN and WDs, are discussed in Chapters 14 and 15, respectively. The formation rate and galactic distribution of PN (Chapter 18) and the testing of the evolutionary models of PN (Chapter 17) are dependent on an accurate knowledge of the distance scale (Chapter 16). The contributions of PN to the chemical structure of galaxies are presented in Chapter 19. The applications of PN as a tool to study the large structure of the universe are discussed in Chapter 20.

2

Ionization structure of planetary nebulae

We begin the discussions on the physics of PN with the classical static model of PN. PN are assumed to be made up of two components: a central star and a surrounding gaseous nebula. If the star is hot enough, much of its energy will be emitted in the ultraviolet (UV). These UV photons will be able to ionize the atoms in the nebula. The electrons ejected in the ionization process provide a pool of kinetic energy for the collisional excitation of the heavy atoms (carbon, nitrogen, oxygen, etc.). Spontaneous emissions from the various excited states of different atoms and ions are responsible for the rich emission-line spectrum seen in the visible.

The first excited state of hydrogen (H) is at 10.2 eV above the ground state, corresponding to an excitation temperature (E/k) of $\sim 10^5$ K. This is much higher than the typical kinetic temperatures of $\sim 10^4$ K found in PN. Even for electrons with energies high enough to overcome this energy gap, the low densities in PN imply that the excitation or ionization rates by electron collisions are much slower than the spontaneous emission rate (see Section 3.4), and the collisionally excited electron will remain at an excited state for a very short time. As the result, the population of an excited state of H is determined not by collisions from below, but by the recombination between free protons and electrons and the subsequent electron cascades *via* spontaneous emissions. The ionization and the excitation states of the H atom can therefore be assumed to be completely controlled by radiation processes.

2.1   Photoionization

The H atom consists of a positively charged nucleus (proton) and an electron. The electron can be in the ground state or any of the excited states. The quantum state of the atom is described by the principal quantum number (n, from 1 to ∞), orbital angular momentum number ($\ell < n$), and the spin quantum number of the electron ($s = \pm 1/2$). The lower case letters s, p, d, f, and so on are used to represent $\ell = 0, 1, 2, 3, \ldots$. For example, the ground state of H is $1s$, and the first excited states of H are $2s$ and $2p$ (Fig. 2.1). Since the only force in the system, the electromagnetic interaction between the proton and the electron, is a central force, all the orbital angular momentum states of the same principal quantum state are degenerate. The energy of the bound electron is a function of n only and is given by the Bohr formula:

$$E_n = h R_{\mathrm{H}} \left(1 - \frac{1}{n^2} \right), \tag{2.1}$$

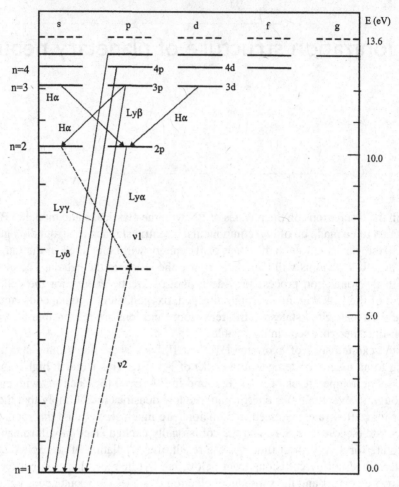

Fig. 2.1. The energy diagram of H atom.

where

$$R_H = \frac{2\pi^2 e^4 m'_e}{h^3} \tag{2.2}$$

is the Rydberg constant for H and m'_e is the reduced electron mass.

If the electron is initially at level $n\ell$ and a photon is absorbed by the atom, the electron can be excited to a higher level $n'\ell'$, or to a free state. The latter process is referred to as photoionization (or bound free). Photoionization will occur if the frequency (ν) of the incoming photon is greater than ν_n:

$$\nu_n = \frac{R_H}{n^2}. \tag{2.3}$$

The Lyman limit ($\nu_1 = 3.3 \times 10^{15}$ Hz) is defined as the minimum frequency required to raise an electron from the ground state of H to a free state. The excess energy of the photon after overcoming the ionization potential will be transferred to the free electron in the form of kinetic energy:

$$\frac{1}{2}m_e v^2 = h\nu - h\nu_n \tag{2.4}$$

One can generalize the Bohr formula to include the continuum states by defining a quantum number k, where $n = ik$. The energy difference between a bound state n and a free state k is then given by

$$E_n = hR_H \left(\frac{1}{n^2} + \frac{1}{k^2}\right). \tag{2.5}$$

Comparing Eqs. (2.4) and (2.5), we have

$$\frac{1}{2}m_e v^2 = \frac{hR_H}{k^2}. \tag{2.6}$$

The cross section $(a_{n\ell})$ for an atom at an initial state $n\ell$ to be ionized by a photon of energy $h\nu$ to a free state of k can be defined by

$$\frac{dP}{dt} = F_\nu n_{n\ell} a_{n\ell}(k^2) d(h\nu), \tag{2.7}$$

where $F_\nu \, d(h\nu)$ is the flux of photons having energies between $h\nu$ and $h\nu + d(h\nu)$, and dP/dt is the number of ionizations per unit volume per unit time. dP/dt can be calculated by the time-dependent perturbation theory and is proportional to the electric dipole moment matrix element. The cross sections for hydrogenic ions of nuclear charge Z can be written in the following form:

$$a_{n\ell}(k^2) = \left(\frac{4\pi\alpha a_0^2}{3}\right) \frac{n^2}{Z^2} \sum_{\ell'=\ell\pm1} \frac{\max(\ell, \ell')}{2\ell + 1} \Theta(n\ell, K\ell'), \tag{2.8}$$

where

$$\Theta(n\ell, K\ell') = (1 + n^2 K^2) \left|g(n\ell, K\ell')\right|^2, \tag{2.9}$$

$$g(n\ell, K\ell') = \frac{Z^2}{n^2} \int_0^\infty R_{n\ell}(r) r F_{k\ell'}(r) dr, \tag{2.10}$$

$$K = \frac{k}{Z^2}, \tag{2.11}$$

a_0 is the Bohr radius, ℓ' is the angular momentum quantum number of the ejected electron, $R_{n\ell}(r)$ and $F_{k\ell'}(r)$ are the initial and final radial wave functions of the ejected electron, respectively; and

$$\alpha = \frac{2\pi e^2}{hc} \tag{2.12}$$

is the fine structure constant.

The integral in Eq. (2.10) can be evaluated numerically and the values of $a_\nu(n\ell)$ are tabulated in Burgess (1965). For historical reasons, the bound-free (b-f) cross section is often expressed as the classical expression of Kramers multiplied by a correction factor known as the Gaunt factor $[g_{bf}(n, \ell)]$:

$$a_\nu(n\ell) = \frac{32}{3\sqrt{3}} \frac{\pi^2 e^6}{ch^3} \frac{R_Z Z^4}{n^5 \nu^3} g_{bf}(n, \ell) \tag{2.13}$$

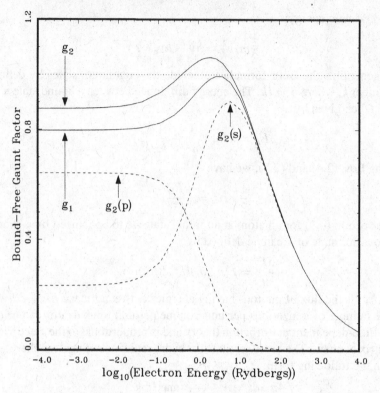

Fig. 2.2. Variation of the bound-free Gaunt factor as a function of frequency. The solid curves show the Gaunt factors for $n = 1$ and 2. The dashed curves show the contributions from the $\ell = 0$ and 1 states to the total $n = 2$ value.

where R_Z is the Rydberg constant for the hydrogenic ion Z. Values for $g_{\mathrm{bf}}(n, \ell)$ can be found in Karzas and Latter (1961). The Gaunt factors for the $n = 1$ and $n = 2$ states of H are plotted in Fig. 2.2. We can see that the Gaunt factor varies slowly with frequency and the frequence dependence is mainly determined by the ν^{-3} term.

Averaging over the angular momentum states and expressing the photon energy in units of ν/ν_n, we have

$$a_n(\nu) = \sum_{\ell=0}^{n-1} \left(\frac{2\ell+1}{n^2} \right) a_{n\ell}$$

$$= \left[\frac{64}{3\sqrt{3}} \alpha \right] (\pi a_0^2) \, n \left(\frac{\nu_n}{\nu} \right)^3 g_{\mathrm{bf}}(n, \nu), \tag{2.14}$$

where

$$g_{n,\mathrm{bf}}(\nu) = \sum_{\ell=0}^{n-1} \left(\frac{2\ell+1}{n^2} \right) g_{\mathrm{bf}}(n\ell). \tag{2.15}$$

The numerical value of the constant in square brackets in Eq. (2.14) is 0.09, so the bound-free cross section is approximately one-tenth of the geometric cross section of the H atom. At the Lyman limit, $g_{\mathrm{bf}}(1s)$ has a value of 0.7973 and the bound-free absorption coefficient for the ground state of H(1s) is 6.3×10^{-18} cm^2. The mean-free path for a photon at the Lyman limit is

$$l \sim \frac{1}{n_{1s} a_{v_1}(1s)}, \tag{2.16}$$

$$\sim \frac{5 \times 10^{-2}}{n_{1s}/\text{cm}^{-3}} \text{ pc}, \tag{2.17}$$

where n_{1s} is the number density of H atoms in the ground state. Assuming that most of H atoms are in the ground state ($n_{1s} = n_H$), a Lyman continuum photon will travel only 5×10^{-5} pc before being absorbed by a H atom in a nebula with $n_H \sim 10^3$ cm^{-3}. Since these distances are much smaller than the typical PN sizes of 0.1 pc, Lyman continuum photons are likely to be trapped inside a dense nebula.[†]

The corresponding optical depth across a nebula length L is

$$\tau = n_{1s} a_{v_1}(1s) L \tag{2.18}$$

$$\sim 19(n_H/\text{cm}^{-3})(L/\text{pc}). \tag{2.19}$$

For a typical PN with $n_H \sim 10 - 10^6$ cm^{-3}, the range for τ is 19 to 1.9×10^6. This suggests that PN are likely to be optically thick in the Lyman continuum. Such nebulae are also referred to as being ionization bounded.

2.2 Recombination

The energy of the free electrons created by photoionization is dependent on the energy of the stellar photons. However, these electrons thermalize very quickly because of the very large electron-electron collisional cross sections. Although low-energy electrons are more likely to recombine with protons [see Eq. (2.29)] and are therefore selectively removed from the free electron pool, the recombination rates [(see Eq. (2.33)] are very slow in comparison with the interactions among electrons themselves (Bohm and Aller, 1947). It is therefore an excellent approximation in the nebular model that the electrons are in local thermodynamic equilibrium (LTE), and their energy distribution can be characterized by a single parameter T_e, the kinetic temperature of the electron.

Since both the photoionization cross section [$a_{n\ell}(v)$] and the recombination cross section ($\sigma_{n\ell}$) are atomic parameters, their values are independent of environmental conditions. We can therefore derive the relationship between these two parameters under the condition of LTE. Under such a condition, the radiation temperature and the kinetic temperature degenerate to the same quantity T. The principle of detailed balance requires the photoionization rate of a neutral atom in excited state $n\ell$ to be equal to the

---

[†] Even for the general interstellar medium where the H density is low, e.g. 1 atom cm^{-3}, the mean free path for an UV photon is much smaller than the interstellar distances. In other words, the interstellar medium is always "cloudy." Although the photoionization cross section is not much different in the visible, the lack of H atoms in an excited state to allow for a bound-free transition makes the mean-free path for visible photons much longer. The fact that we can see stars at night is a testimony to the assumption that very few H atoms in the interstellar medium are in an excited state.

recombination rate by electron capture:

$$4\pi n_{n\ell} a_\nu(n\ell) B_\nu(T)(1 - e^{-h\nu/kT})\frac{d\nu}{h\nu} = n_p n_e \sigma_{n\ell}(v) f(v, T)v\,dv, \tag{2.20}$$

where $n_{n\ell}$ is the number of H atoms in excited state (n, ℓ), $\sigma_{n\ell}$ is the recombination cross section from a free state to level (n, ℓ), B_ν is the Planck function, and $f(v)$ is the Maxwellian distribution of electron velocity at temperature T_e:

$$f(v) = \frac{4}{\sqrt{\pi}}\left(\frac{m_e}{2kT}\right)^{3/2} v^2 e^{-m_e v^2/2kT}. \tag{2.21}$$

The factor $(1 - e^{-h\nu/kT})$ in Eq. (2.20) is to account for the effects of stimulated emission in LTE. Under LTE conditions, the number density of neutral H atoms in the ground state $(1s)$ is related to the proton (n_p) and the electron density (n_e) by the Saha equation:

$$\frac{n_{1s}}{n_p n_e} = \left(\frac{h^2}{2\pi m_e kT}\right)^{3/2} e^{h\nu_1/kT}. \tag{2.22}$$

The population in an excited state $n\ell$ is given by the Boltzmann equation:

$$n_{n\ell} = n_{1s}(2\ell + 1)e^{-E_n/kT}. \tag{2.23}$$

Combining Eqs. (2.22) and (2.23), we have:

$$n_{n\ell} = n_p n_e(2\ell + 1)\left(\frac{h^2}{2\pi m_e kT}\right)^{3/2} e^{h\nu_n/kT}. \tag{2.24}$$

Substituting Eqs. (2.24) and (2.21) into Eq. (2.20), and making use of the relationships

$$h\nu = h\nu_n + \frac{1}{2}m_e v^2, \tag{2.25}$$

$$h\,d\nu = m_e v\,dv, \tag{2.26}$$

we have

$$\frac{a_\nu(n\ell)}{\sigma_{n\ell}(v)} = \frac{1}{2(2\ell + 1)}\frac{m_e^2 v^2 c^2}{h^2 v^2}. \tag{2.27}$$

Summing over all orbital angular momentum states, we have

$$\frac{a_\nu(n)}{\sigma_n(v)} = \frac{m_e^2 v^2 c^2}{h^2 v^2}\frac{1}{2n^2}. \tag{2.28}$$

Substituting Eq. (2.13) into Eq. (2.28), we have

$$\sigma_n(v) = \left(\frac{16}{3\sqrt{3}}\frac{e^2 h}{m_e^2 c^3}\right)\left(\frac{\nu_1}{\nu}\right)\left(\frac{h\nu_1}{\frac{1}{2}m_e v^2}\right)\left(\frac{g_{bf}}{n^3}\right). \tag{2.29}$$

We note that this relationship is true whether the atom is in LTE or not. The constants in the first parentheses have a numerical value of 2.105×10^{-22} cm^2. Since the other three terms in parentheses are all of the order of unity, the recombination cross sections are therefore much smaller than the geometric cross section of the H atom.

The recombination coefficient to level $n\ell$ from an ensemble of electrons at temperature T_e is given by

$$\alpha_{n\ell}(T_e) = \int_0^\infty \sigma_{n\ell} v f(v) dv, \qquad (2.30)$$

and the total recombination coefficient is the sum over captures to all states:

$$\alpha_A = \sum_{n=1}^\infty \sum_{\ell=0}^{n-1} \alpha_{n\ell}(T_e). \qquad (2.31)$$

Values of the recombination coefficients $\alpha_{n\ell}$ are tabulated in Burgess (1965). For $T_e = 10^4$ K, α_A for H has a value of 4.18×10^{-13} cm^3 s^{-1}. The time scale for recombination can be estimated by

$$t_r = \frac{1}{n_e \alpha_A} \qquad (2.32)$$

$$\sim \frac{7.6 \times 10^4}{(n_e / \text{cm}^{-3})} \text{ yr.} \qquad (2.33)$$

The recombination time scale is relatively short under high nebular density (e.g., when the PN is young), but it can be comparable to the dynamical time scale at the later stages of evolution of PN (see Section 13.3).

Assuming that the recombination time is short compared to the dynamical time, the nebula can be assumed to be static. Since the recombination rate is much slower than the spontaneous decay rate (see Section 3.4), the H atom will quickly cascade to the ground state after each recombination and practically all neutral H atoms can be assumed to be in the ground state. At each point in the nebula, the photoionization rate from the ground state is balanced by the total recombination rate to all levels of the H atom,

$$n_{1s} \int_{\nu_1}^\infty \frac{4\pi J_\nu}{h\nu} a_\nu(1s) d\nu = n_p n_e \alpha_A(T_e) \qquad (2.34)$$

where $J_\nu = \int I_\nu d\Omega / 4\pi$ is the mean radiation intensity at that point.

However, every recombination to the ground state will create another Lyman continuum photon, which can be absorbed by another nearby H atom. If the nebula is ionization bounded (i.e. no Lyman continuum photons escape), the recombination to the ground state has no net effect on the overall ionization balance of the nebula. Since the mean-free path of these diffuse Lyman continuum photons is short, see Eq. (2.17), their existence can be ignored by not counting the ground state in the recombination process. In other words, the total number of ionizing photons emitted by the star of temperature T_* and radius R_*,

$$Q = \int_{\nu_1}^\infty \frac{4\pi R_*^2 \pi B_\nu(T_*)}{h\nu} d\nu, \qquad (2.35)$$

should be balanced by the total number of recombinations to excited states ($n > 1$) within the ionized volume,

$$Q = \int n_p n_e \alpha_B dV, \qquad (2.36)$$

where

$$\alpha_B = \sum_{n=2}^{\infty} \sum_{\ell=0}^{n-1} \alpha_{n\ell}(T_e).$$ (2.37)

At $T_e = 10,000$ K, α_B has a value of 2.59×10^{-13} cm^3 s^{-1}. Solving the equation of transfer together with the ionization equilibrium equation [Eq. (2.34)] shows that the H atom is nearly completely ionized out to a critical radius r_s, at which the ionization state changes to neutral over a very short distance. In other words, inside r_s, $n_p = n_e = n_H$, and outside r_s, $n_p = n_e = 0$. In this case, Eq. (2.36) can be integrated to give

$$Q = \frac{4\pi}{3} r_s^3 n_H^2 \alpha_B,$$ (2.38)

and r_s is referred to as the Strömgren radius (Strömgren, 1939).

Equation (2.35) can be written as

$$Q = \frac{8\pi^2 R_*^2}{c^2} \left(\frac{kT_*}{h} \right)^3 G(T_*),$$ (2.39)

where

$$G(T_*) = \int_{h\nu_1/kT_*}^{\infty} \frac{x^2}{e^x - 1} dx.$$ (2.40)

and $x = h\nu/kT_*$. If the star radiates like a blackbody, the total stellar luminosity is given by

$$L_* = 4\pi R_*^2 \sigma T_*^4,$$ (2.41)

and Eq. (2.39) can be expressed in terms of the stellar luminosity

$$Q = \frac{15G(T_*)}{\pi^4 kT_*} L_*.$$ (2.42)

The Strömgren radius is then given by

$$r_s = \left[\frac{45 G(T_*) L_*}{4\pi^5 k T_* n_H^2 \alpha_B} \right]^{\frac{1}{3}}.$$ (2.43)

Figure 2.3 shows the Strömgren radius as a function of T_* for a star with $L_* = 10^4\, L_\odot$ and three different nebular densities.

2.3   Ionization structure of a static nebula

Assuming that the nebular dynamical time and the evolutionary time for the central star are long compared to the recombination time, the ionization structure of the nebula can be calculated for He and heavy elements. An atom X of atomic number N has $N + 1$ ionization states, and the abundance of each ion (X^i) in each part of the nebula

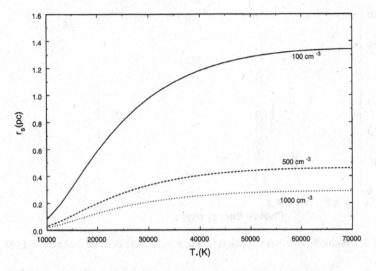

Fig. 2.3. The Strömgren radius as a function of stellar temperature for a star of luminosity $10^4 \, L_\odot$ embedded in a nebula of densities 100, 500, and 1000 cm^{-3}.

is determined by the balance of photoionization and recombination from the next $(i + 1)$ ionization state,

$$N_1(X^i) \int_{\nu_1}^{\infty} \frac{4\pi J_\nu a_1(\nu, X^i)}{h\nu} d\nu$$

$$= \sum_{j+1}^{\infty} N_j(X^{i+1}) n_e \times \int_0^{\infty} f(\upsilon) \upsilon \alpha_j(\upsilon, X^i) d\upsilon, \qquad (2.44)$$

where N_j is the number density of the ion in the jth excitation state ($j = 1$ is the ground state). Here we have assumed the rate of collisional ionization and excitation to be negligible and that all photoionizations occur from the ground state.

Since each emission line we observe comes from one ionization state of an atom, an accurate ionization model is necessary to correctly deduce the total elemental abundance from the strength of one line. This requires a good knowledge of the photoionization and recombination coefficients, which can be difficult to obtain for complex atoms.

2.4 Ionization of complex atoms

The electron configurations of multi-electron atoms under spin-orbit (LS) coupling can be conveniently expressed in the spectroscopic notation $^{2S+1}L_J$, where S, L, and J are the spin, orbit, and total angular momentum quantum numbers, respectively. The term $2S + 1$ is referred to as the multiplicity. For example, the He atom can be in the $S = 0$ (antiparallel electron spin) or $S = 1$ (parallel electron spin) states, corresponding to multiplicity values of 1 (singlet) and 3 (triplets) respectively. The permissible values of J determined by vector coupling of L and S are $L + S, L + S - 1, \ldots, |L - S|$. For example, the states of the p^2 (such as the ground state of C, $1s^2 2s^2 2p^2$) configuration are $^3P_{2,1,0}$, $^1D_2$, and $^1S_0$. The parity of a multielectron atomic state is determined by $(-1)^{\Sigma \ell}$.

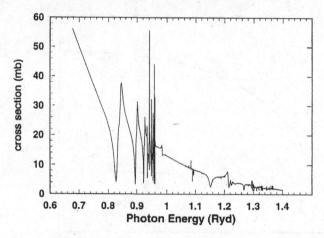

Fig. 2.4. Photoionization cross section for the ground state of Si (from Ferland, 1993).

The parity (odd or even) of the state is sometimes added as a superscript (e.g. the lowest energy term of N with a configuration of $1s^2 2s^2 2p^3$ is written as $^4 S^o_{3/2}$).

In addition to H, photoionization can also occur from the ground state of more complex atoms. For example, NI can be photoionized to a number of states of NII

$$
\text{NI}\,(2s^2 2p^3\,{}^4 S_{3/2}) + h\nu \rightarrow
\begin{array}{l}
\text{NII}\,(2s^2 2p^2\,{}^3 P) + e^- h\nu > 14.5\,\text{eV} \\
\text{NII}\,(2s^2 2p^2\,{}^1 D) + e^- h\nu > 16.4\,\text{eV} \\
\text{NII}\,(2s^2 2p^2\,{}^1 S) + e^- h\nu > 19.6\,\text{eV}.
\end{array}
\tag{2.45}
$$

The cross sections at the photoionization threshold for the above three processes are $1.8 \times 10^{-18}, 2.5 \times 10^{-18}$, and 1.5×10^{-18} cm^2 respectively. Although the photoionization cross section of H has the simple ν^{-3} dependence, a_ν of complex atoms have more complicated structures (Fig. 2.4). The presence of resonance features suggests that an ion can make significant contributions to the optical depth of the nebula at specific frequencies in spite of the low abundance of the ion.

2.5   Dielectric recombination

The recombination of multielectron ions is much more complicated than that of H because of the presence of other bound electrons. For example, the direct capture of a free electron to the innermost unfilled electron shell has to satisfy Pauli's exclusion principle. The recombination coefficient can be much higher at certain energies (called resonances) if the energy of the incoming electron corresponds exactly to the energy of a doubly excited state of a bound level of a lower ionization state of an ion. This ion can then radiatively decay to a singly excited state and then to the ground level.

As an example, the main recombination for doubly ionized C follows the process

$$
\text{CIII}\,(2s^2) + e(0.41\,\text{eV}) \rightarrow \text{CII}\,(2s2p, 3d),
\tag{2.46}
$$
$$
\rightarrow \text{CII}\,(2s2p^2) + h\nu,
\tag{2.47}
$$
$$
\rightarrow \text{CII}\,(2s^2 2p) + h\nu,
\tag{2.48}
$$

where the free electron excites one of the bound electrons to the $2p$ state while it settles into the $3d$ shell. The resultant singly ionized C then undergoes two separate decays to return to the ground state. This process is referred to as dielectronic recombination (Storey, 1981).

The total recombination coefficients, including the effects of both radiative and dielectronic recombinations, for multielectron ions such as Si and C have been tabulated by Nahar (1995).

2.6   Charge-exchange reactions

The recombination of certain doubly and more highly charged ions can take place by means of charge exchange with neutral H or He:

$$A^{+q} + H \rightarrow A^{+q-1} + H^+. \tag{2.49}$$

Usually the ion recombines to an excited state and decays rapidly by spontaneous emission. Consequently, the reverse process is not important. An example of the charge-exchange reaction is

$$N^{+3}(2s^2)\,^1S + H(1s)\,^2S \rightarrow N^{+2}(2s^2 3s)\,^2S + H^+, \tag{2.50}$$

which has a rate of 2.93×10^{-9} cm^3 s^{-1} at $T_e = 10^4$ K. Sometimes the electron captured to the valence orbital is accompanied by a rearrangement of the core orbitals:

$$N^{++}(2s^2 2p)\,^2P + H(1s)\,^2S \rightarrow N^+(2s2p^3)\,^3D + H^+, \tag{2.51}$$

which has a rate of 0.86×10^{-9} cm^3 s^{-1} at $T_e = 10^4$ K.

In ionization equilibrium models, the charge-exchange rates can simply be added to the photoionization and recombination terms.

3

Nebular line radiation

Unlike stars which show a continuous spectrum, the optical spectrum of PN is dominated by emission lines. Line emission occurs when atoms or ions make a transition from one bound electronic state to another bound state at a lower energy. Such transitions, usually by means of spontaneous emission, are referred to as bound-bound (b-b) transitions. In the interior of stars, electrons in an atom are distributed over many energy levels because of the high particle and radiation densities. The bound electrons are excited either by free electrons colliding with the atom, or by the absorption of a photon. However, in the interstellar medium, both the particle and radiation densities are low, and the population distribution of the bound electrons can be far from the thermodynamical equilibrium condition given by the Boltzmann equation [Eq. (2.23)].

The typical energy separations between the electronic states of atoms are of the order of 1 eV, corresponding to photons in the visible or UV parts of the spectrum. The only available visible or UV background in the interstellar medium is from diluted starlight, which is generally not strong enough for excitation by stimulated absorption to be significant. Therefore the only way that a bound electron can be found in an excited state is by collisional excitation from a lower state, or as a consequence of recombination between a free electron and a proton. The line photons emitted as the result of collisional excitation are called collisionally excited lines, and those emitted following recombination are called recombination lines. H and He, with large energy gaps between the first excited state and the ground state, are difficult to excite by collisions, whereas heavy atoms, with their more complicated electronic structures, often have low-lying electronic states within fractions of an electron volt from the ground state and can be more effectively excited by collisions.

3.1 Permitted and forbidden transitions

The selection rules in LS coupling are $\Delta L = \pm 1$ or 0, $\Delta S = 0$, and $\Delta J = 0, \pm 1$ except $J = 0 \rightarrow 0$. Since the emitted photon carries angular momentum, this requires at least one electron to change its angular momentum ($\Delta \ell = \pm 1$, which is equivalent to the parity selection rule of even $\leftrightarrow$ odd). The $\Delta S = 0$ rule implies that transitions will occur only between terms of the same multiplicity. In the case of He, which is in pure LS coupling, transitions will occur only from a singlet state to a singlet state or a triplet state to a triplet state, and the singlet and triplet states of He can be regarded as two separate atoms. Transitions from singlet to triplet states of ions of light elements such

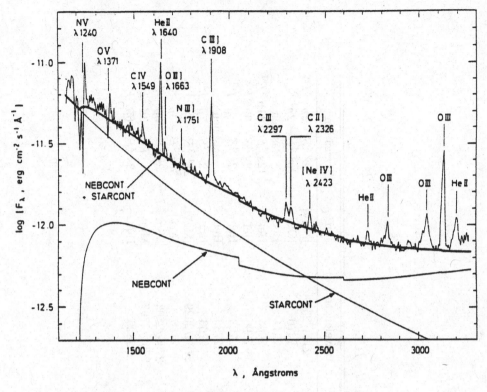

Fig. 3.1. *IUE* spectrum of NGC 7009 (from Harrington *et al.*, 1981).

as CIII, OIII, and NIV (e.g. $1s^2 2s^2\ {}^1S - 1s^2 2s2p\ {}^3P$ at λ 1,906 Å, 1,908 Å of CIII) are referred to as intercombination transitions and are only semiforbidden because the ions are only close to LS coupling. These lines are indicated by a bracket on the right-hand side (e.g., NIII]).

The strongly allowed transitions to the ground state (e.g., the ${}^2P_{3/2} - {}^2S_{1/2}\ \lambda$ 5,890 Å and ${}^2P_{1/2} - {}^2S_{1/2}\ \lambda$ 5,896 lines of NaI) are known as resonance lines. Permitted and intercombination lines of C, N, O, and Si dominate the near-ultraviolet spectrum of PN (Fig. 3.1). A list of lines observed in PN by the *International Ultraviolet Explorer (IUE)* is given in Table 3.1.

Transitions that violate the LS coupling selection rules can still occur as electric quadrupole or magnetic dipole transitions, but their spontaneous emission coefficients (A_{ij}) are much smaller than those of permitted electric dipole transitions. In terrestrial conditions, the high frequency of collisions means that the upper state will be collisionally deexcited before it has a chance to self-decay. However, in interstellar conditions, the time between collisions can be longer than the lifetime of the upper state, allowing the upper state to decay by emitting a forbidden-line photon.

Many of the strongest lines from PN are forbidden lines. Figure 3.2 shows two examples (OII and SII) under the p^3 configuration. All transitions shown in Fig. 3.2 are forbidden because they involve transitions within the same configuration and therefore violate the $\Delta\ell = \pm1$ selection rule. These lines are indicated by square brackets (e.g., [OII]).

Table 3.1. *Atomic lines in PN spectra detected by the IUE*

Wavelength (Å)	Ion	Transition
1175/76	CIII	$2p\,^3P^0 - 2p^2\,^3P$
1239/43	NV	$2s\,^2S - 2p\,^2P^0$
1309	SiIII	$3p\,^2P^0 - 3p^2\,^2S$
1335/36	CII	$2p\,^2P^0 - 2p^2\,^2D$
1371	OV	$2p\,^1P^0 - 2p^2\,^1D$
1394/1403	SiIV	$3s\,^2S - 3p\,^2P^0$
1397–1407	OIV]	$2p\,^2P^0 - 2p^2\,^4P$
1483/87	NIV]	$2s^2\,^1S - 2p\,^3P^0$
1548/50	CIV	$2s\,^2S - 2p\,^2P^0$
1575	[NeV]	$2p^2\,^3P - 2p^2\,^1S$
1602	[NeIV]	$2p^3\,^4S^0 - 2p^3\,^2P^0$
1640	HeII	Balmer α
1658–66	OIII]	$2p^2\,^3P - 2p^3\,^5P^0$
2423/25	[NeIV]	$2p^3\,^4S - 2p^3\,^2D$
2470	[OII]	$2p^3\,^4S - 2p^3\,^2P$
2511	HeII	Paschen γ
2663	HeI	$2s\,^3S - 11p\,^3P^0$
2696	HeI	$2s\,^3S - 9p\,^3P^0$
2723	HeI	$2s\,^3S - 8p\,^3P^0$
2733	HeII	Paschen β
2763	HeI	$2s\,^3S - 7p\,^3P^0$
2784/2929	[MgV]	$2p^4\,^3P - 2p^4\,^1D$
2786	[ArV]	$3p^2\,^3P - 3p^2\,^1S$
2796/2803	MgII	$3s\,^2S - 3p\,^2P^0$
2791/2797	MgII	$3p\,^2P^0 - 3d\,^2D$
2829	HeI	$2s\,^3S - 6p\,^3P^0$

1718	NIV	$2p\,^1P^0 - 2p^2\,^1D$	2837/38	CII	$2p^2\,^2S - 3p\,^2P^0$
1711	SiII	$3p^2\,^2D - 5f\,^2F^0$	2837	OIII	$3p\,^3D - 3d\,^3P^0$
1747–54	NIII]	$2p\,^2P^0 - 2p^2\,^4P$	2852	MgI	$3s^2\,^1S - 3p\,^1P^0$
1760	CII	$2p^2\,^2D - 3p\,^2P^0$	2854/68	[ArIV]	$3p^2\,^4S - 3p^2\,^2P$
1815	[NeIII]	$2p^4\,^3P - 2p^4\,^1S$	2929/37	MgII	$3p\,^2P^0 - 4s\,^2S$
1808/17	SiII	$3p\,^2P^0 - 3p^2\,^2D$	2929	[MgV]	$2p^4\,^3P - 2p^4\,^1D$
1882/92	SiIII]	$3s^2\,^1S - 3p\,^3P^0$	2945	HeI	$2s\,^3S - 5p\,^3P^0$
1907/09	CIII]	$2s^2\,^1S - 2p\,^3P^0$	2973/79	NIII	$3p\,^2P^0 - 3d\,^2P^0$
2253	HeII	Paschen 6	3023	OIII	$3s\,^3P^0 - 3p\,^3P$
2297	CIII	$2p\,^1P^0 - 2p^2\,^1D$	3043/47	OIII	$3s\,^3P^0 - 3p\,^3P$
2306	HeII	Paschen ϵ	3063/71	[NII]	$2p^2\,^3P - 2p^2\,^1S$
2321/31	[OIII]	$2p^2\,^3P - 2p^2\,^1S$	3109/3005	[ArIII]	$3p^4\,^3P - 3p^4\,^1S$
2325–29	CII]	$2p\,^2P^0 - 2p^2\,^4P$	3133	OIII	$3p^3S - 3d\,^3P^0$
2334–50	SiII	$3p\,^2P^0 - 3p^2\,^4P$	3188	HeI	$2s\,^3S - 4p\,^3P^0$
2385	HeII	Paschen δ	3203	HeII	Paschen α

Table taken from Köppen and Aller (1987).

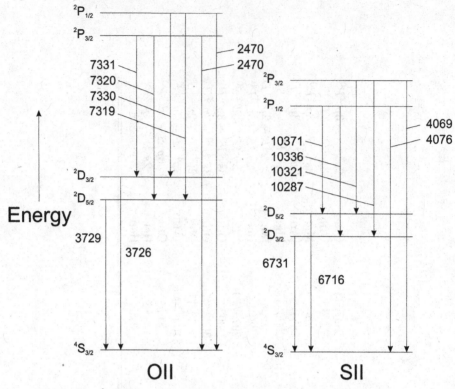

Fig. 3.2. The p^3 electron configuration.

Transitions between states of the same multiplets (splitting caused by spin-orbit inter-actions) are known as fine-structure lines. Examples are the $^3P_{3/2} - ^2P_{1/2}$ line of CII at 158 μm and the $^3P_2 - ^3P_1$ line of CI at 370 μm. Many of these lines are now detected by the *Infrared Space Observatory (ISO)* (see Table 3.2).

3.2   Absorption and emission

In a bound-bound transition, the line absorption coefficient is given by

$$\kappa_\nu = \frac{h\nu}{4\pi}(n_j B_{ji} - n_i B_{ij})\phi_\nu,\tag{3.1}$$

where ϕ_ν is the normalized profile of the absorption line, n_j and n_i are the number densities of the atom in the lower (j) and upper (i) states, and B_{ji} and B_{ij} are the stimulated absorption and emission coefficients, respectively. The emission coefficient is given by

$$j_\nu = \frac{h\nu}{4\pi}n_i A_{ij}\phi_\nu,\tag{3.2}$$

Table 3.2. *Atomic and Ionic Fine-Structure Lines in the Infrared*

Ion	Transition	Wavelength (μm)	E (eV)	I.P (eV)
[SiVII]	$^3P_1 - {}^3P_2$	2.483	205.05	246.52
[AlV]	$^2P_{1/2} - {}^2P_{3/2}$	2.905	120.00	153.83
[CaIV]	$^2P_{1/2} - {}^2P_{3/2}$	3.207	50.91	67.27
[MgIV]	$^2P_{1/2} - {}^2P_{3/2}$	4.488	80.14	109.24
[ArVI]	$^2P_{3/2} - {}^2P_{1/2}$	4.528	75.02	91.01
[MgV]	$^3P_1 - {}^3P_2$	5.608	109.24	141.27
[KIV]	$^3P_1 - {}^3P_2$	5.983	45.80	60.91
[ArII]	$^2P_{1/2} - {}^2P_{3/2}$	6.985	15.76	27.63
[NaIII]	$^2P_{1/2} - {}^2P_{3/2}$	7.319	47.29	71.62
[ArV]	$^3P_2 - {}^3P_1$	7.902	59.81	75.02
[NaIV]	$^3P_1 - {}^3P_2$	9.041	71.62	98.91
[SIV]	$^2P_{3/2} - {}^2P_{1/2}$	10.510	34.79	47.22
[CaV]	$^3P_0 - {}^3P_1$	11.482	67.27	84.50
[NiII]	$^4F_{5/2} - {}^4F_{7/2}$	12.729	7.64	18.17
[NeII]	$^2P_{1/2} - {}^2P_{3/2}$	12.814	21.56	40.96
[ArV]	$^3P_1 - {}^3P_0$	13.102	59.81	75.02
[MgV]	$^3P_0 - {}^3P_1$	13.521	109.24	141.27
[NeV]	$^3P_2 - {}^3P_1$	14.320	97.12	126.21
[NeIII]	$^3P_1 - {}^3P_2$	15.555	40.96	63.45
[SIII]	$^3P_2 - {}^3P_1$	18.713	23.34	34.79
[ArIII]	$^3P_0 - {}^3P_1$	21.842	27.63	40.74
[NeV]	$^3P_1 - {}^3P_0$	24.316	97.12	126.21
[OIV]	$^2P_{3/2} - {}^2P_{1/2}$	25.913	54.93	77.41
[SIII]	$^3P_1 - {}^3P_0$	33.480	23.34	34.79
[SiII]	$^2P_{3/2} - {}^2P_{1/2}$	34.814	8.15	16.35
[NeIII]	$^3P_0 - {}^3P_1$	36.009	40.96	63.45
[FIV]	$^3P_1 - {}^3P_0$	44.070	62.71	87.14
[FeIII]	$^5D_1 - {}^5D_2$	51.680	16.19	30.65
[OIII]	$^3P_2 - {}^3P_1$	51.814	35.12	54.93
[SI]	$^3P_0 - {}^3P_1$	56.311	0.00	10.36
[NIII]	$^2P_{3/2} - {}^2P_{1/2}$	57.317	29.60	47.45
[PII]	$^3P_1 - {}^3P_0$	60.640	10.48	19.77
[OI]	$^3P_1 - {}^3P_2$	63.184	0.00	13.62
[FII]	$^3P_0 - {}^3P_1$	67.200	17.42	34.97
[SiI]	$^3P_2 - {}^3P_1$	68.473	0.00	8.15
[OIII]	$^3P_1 - {}^3P_0$	88.356	35.12	54.93
[AlI]	$^2P_{3/2} - {}^2P_{1/2}$	89.237	0.00	5.99
[FeIII]	$^5D_0 - {}^5D_1$	105.370	16.19	30.65
[NII]	$^3P_2 - {}^3P_1$	121.898	14.53	29.60
[SiI]	$^3P_1 - {}^3P_0$	129.682	0.00	8.15
[OI]	$^3P_0 - {}^3P_1$	145.525	0.00	13.62
[CII]	$^2P_{3/2} - {}^2P_{1/2}$	157.741	11.26	24.38
[NII]	$^3P_1 - {}^3P_0$	205.178	14.53	29.60

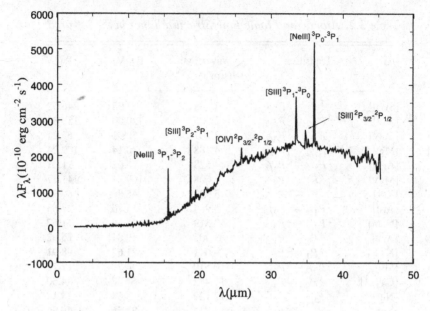

Fig. 3.3. ISO SWS spectrum of H1-12 showing several fine-structure lines. The continuum is due to dust emission (see Section 6.2).

The source function $S_\nu = j_\nu / \kappa_\nu$ is therefore

$$S_\nu = \frac{A_{ij}/B_{ij}}{\frac{n_j B_{ji}}{n_i B_{ij}} - 1}. \qquad (3.3)$$

Under LTE, the source function is given by the Planck function (Kirchoff's law)

$$S_\nu = B_\nu = \frac{2h\nu^3}{c^2} \frac{1}{e^{h\nu/kT} - 1}, \qquad (3.4)$$

and the energy distribution of an atom is given by the Boltzmann equation

$$\frac{n_i}{n_j} = \frac{g_i}{g_j} e^{-h\nu/kT}, \qquad (3.5)$$

where g_i and g_j are the degrees of degeneracy in the lower and upper states respectively. Combining Eqs. (3.3), (3.4), and (3.5), we have the following identities:

$$\frac{g_j}{g_i} \frac{B_{ji}}{B_{ij}} = 1, \qquad \frac{A_{ij}}{B_{ij}} = \frac{2h\nu^3}{c^2}. \qquad (3.6)$$

Since the Einstein A and B coefficients are fundamental properties of the atom and are not dependent on the environmental conditions, these identities are true even if the atom is not in LTE.

Substituting Eq. (3.6) into Eq. (3.3), we can express the source function as

$$S_\nu = \frac{2h\nu^3/c^2}{\frac{g_i n_j}{g_j n_i} - 1}. \qquad (3.7)$$

We can define the atomic absorption cross section a_ν by

$$a_\nu = \frac{\kappa_\nu}{n_j}. \tag{3.8}$$

Comparing this equation with Eq. (3.1) and making use of Eq. (3.6), we have

$$a_\nu = \left(1 - \frac{n_i}{n_j}\frac{g_j}{g_i}\right)\frac{h\nu}{4\pi}B_{ji}\phi_\nu, \tag{3.9}$$

The oscillator strength of a line is defined by

$$f_{ji} = \frac{m_e c}{\pi e^2}\frac{h\nu}{4\pi}B_{ji}, \tag{3.10}$$

Substituting Eq. (3.10) into Eq. (3.6), we have

$$A_{ij} = \frac{g_j}{g_i}\frac{8\pi^2 e^2 \nu^2}{m_e c^3}f_{ji}. \tag{3.11}$$

The absorption cross section can be expressed in terms of the oscillator strength by

$$a_\nu = \left(1 - \frac{g_j n_i}{g_i n_j}\right)\frac{\pi e^2}{m_e c}f_{ji}\phi_\nu \tag{3.12}$$

For atomic transitions, the oscillator strengths can either be calculated or measured. Once they are known, the macroscopic concepts such as the absorption coefficient and optical depth can be obtained through these equations.

3.3 Thermodynamic equilibrium versus steady state

In the interstellar environment, unlike in the terrestrial environment, the energy distributions of free particles (e.g., electrons in the nebula), atoms and molecules, and the radiation field often cannot be characterized by a single temperature. For example, if the frequency of collisions among electrons themselves is high enough, the energy distribution of the electrons can be described by the Maxwellian distribution characterized by T_e. If the frequency of electron-atom collisions is so high than they dominate over all radiative processes, then the population distribution within the atoms can be described by the Boltzmann equation at temperature T_e [Eq. (2.23)]. However, if the spontaneous emission rate is much faster than the collisional rates (e.g., in the case of interstellar H), the population distribution within an atom is not related to T_e. To emphasize this distinction, we use the three different terms: electron kinetic temperature (T_e), brightness temperature (T_b), and excitation temperature (T_x) to describe the behavior of free particles, the radiation field, and quantum systems (see Chapter 5).

Instead of the assumption of LTE, we can assume that the nebula is in a steady state where the population distribution is not changing with time. In this case, the population at each level is governed by the statistical equilibrium equation where the incoming rate to each level i is exactly balanced by the outgoing rate:

$$n_j\left[\sum_k (B_{jk}\bar{J}_{jk} + n_e C_{jk}) + \sum_{k<j} A_{jk}\right]$$

$$= \sum_k n_k(n_e C_{kj} + B_{kj}\bar{J}_{kj}) + \sum_{k>j} A_{kj}, \tag{3.13}$$

where

$$\bar{J}_{ij} = \int I_\nu \phi_\nu d\nu \frac{d\Omega}{4\pi}, \qquad (3.14)$$

and $n_e C_{ji}$ and $n_e C_{ij}$ are the collisional excitation and deexcitation rates respectively.

The solution to the statistical equilibrium equation can be substituted into Eq. (3.7) to obtain the source function for each transition. This can in turn be substituted into the equation of transfer,

$$\frac{dI_\nu}{d\tau_\nu} = -I_\nu + S_\nu, \qquad (3.15)$$

to solve the intensity (I_ν) as a function of the optical depth.

3.4   Recombination lines

Since the H atom cannot be excited by collisional means under nebular conditions, and there is also very little background interstellar radiation in the visible region available for excitation by stimulated absorption, the only way that an H atom can be found in an excited state is by recombination. After each direct recombination, the atom will cascade to lower states by a series of spontaneous emissions. These lines are called recombination lines. Several examples of H recombination lines are shown in Fig. 2.1.

The spontaneous transition probability $A_{n\ell,n'\ell'}$ for electric dipole transition in hydrogenic ions is given by

$$A_{n\ell,n'\ell'} = \frac{64\pi^4 \nu^3}{Z^2 3hc^3} \frac{\max(\ell, \ell')}{2\ell + 1} e^2 a_0^2 \left[\int_0^\infty R(n'\ell') r \, R(n\ell) dr \right]^2, \qquad (3.16)$$

$$\nu = R_Z Z^2 \left(\frac{1}{n^2} - \frac{1}{n'^2} \right). \qquad (3.17)$$

For electric dipole transitions, the selection rule is $\ell = \ell' \pm 1$.

The values of the spontaneous decay rates (expressed as oscillator strengths) for several lower transitions of H are given in Table 3.3. The oscillator strength of the Lyα line (the first entry in the table) is 0.4162, implying a spontaneous decay rate ($A_{2p,1s}$) of 6.2×10^8 s^{-1}. Since this rate is much faster than the recombination rate, any recombination will quickly decay to the ground state. For practical purposes, the H

Table 3.3. *Oscillator strengths for some of the lower transitions of* H

n_1, ℓ_1	n_2, ℓ_2	f	n_1, ℓ_1	n_2, ℓ_2	f
1,0	2,1	0.4162	2,1	3,0	0.03159
	3,1	0.07910		3,2	0.6958
	4,1	0.02899		4,0	0.003045
	5,1	0.01394		4,2	0.1218
2,0	3,1	0.4349		5,0	0.001213
	4,1	0.1028		5,2	0.04437

atoms can be assumed to be all in the ground state, and the second term in the absorption cross section in Eq. (3.9) can be ignored. Assuming thermal broadening at a temperature of $\sim 10^4$ K, the absorption cross section for the Lyα line is $\sim 10^{-13}$ cm^2. The mean free path for a Lyα photon, given by $1/n_H a_\nu$, is extremely short. A Lyα photon can therefore expect to travel only a small fraction of the nebular size before being absorbed by another H atom. The corresponding optical depth in the line is

$$\tau = \int \int n_1 a_\nu \phi_\nu d\nu dl \sim 3 \times 10^5 \left(\frac{n_H}{cm^{-3}} \right) (L/pc). \qquad (3.18)$$

A comparison with Eq. (2.19) shows that the Lyα line optical depth is 10^4 times the optical depth at the Lyman limit for continuum photons. In other words, if the nebula is ionization bounded, the optical depth of Lyα is $> 10^4$.

Since the Lyman series is optically thick, the level population of the H atoms is affected by stimulated absorption, and a full radiative transfer calculation is needed to determine the population distribution. However, if we take the extreme case that the entire Lyman series is optically thick, then the net effect is that every Lyman series photon is eventually converted to photons of a lower H series (e.g., Balmer, Paschen) plus a Lyα photon. Take the example of Lyβ. After it is emitted, it will be reabsorbed by another H atom until the atom takes the route of cascading to the $n = 2$ state (by emitting a Hβ photon) and then to the $n = 1$ by emitting a Lyα photon. Similarly, a Lyγ photon will be converted to P$\alpha +$ H$\alpha +$ Lyα after several scatterings. The net effect is equivalent to the total absence of the $n = 1$ state. The Lyα photon will be scattered from atom to atom until it escapes from the nebula or is absorbed by dust, but it will have no effect on the population distribution of the remaining states. This approximation (called Case B) allows for the solution to the population distribution by including only spontaneous emission and recombination in the statistical equilibrium equation (collisions are not important as we noted before). The other approximation (called Case A) corresponds to the case in which all the H transitions (including the Lyman series) are optically thin.

Under Case A or Case B, the stimulated absorption/emission terms in the statistical equilibrium equation [Eq. (3.13)] can be omitted. For each level n there are only three terms in the equation: direct capture of a free electron, cascade from levels above, and cascade to levels below:

$$\sum_{i=n+1}^{\infty} \sum_{\ell'} n_{i\ell'} A_{i\ell',n\ell} + n_p n_e \alpha_{n\ell}(T_e) = n_{n\ell} \sum_{j=1or2}^{n-1} \sum_{\ell''} A_{n\ell,j\ell''}, \qquad (3.19)$$

where the sum on the right-hand side begins with 1 for Case A and 2 for Case B. If the population distribution is in LTE, the population in each excited state $n\ell$ will be given by Eq. (2.22). In general, the actual population can be written as

$$n_{n\ell} = b_{n\ell} n_p n_e (2\ell + 1) \left(\frac{h^2}{2\pi m_e k T_e} \right)^{3/2} e^{h\nu_n/kT_e}, \qquad (3.20)$$

where the coefficients $b_{n\ell}$ represent the degree of departure from the thermodynamic equilibrium distribution.

Substituting Eq. (3.20) into Eq. (3.19), we have

$$\frac{\alpha_{n\ell}}{2\ell+1}\left(\frac{2\pi m_e kT_e}{h^2}\right)^{3/2} e^{-h\nu_n/kT_e} + \sum_{i=n+1}^{\infty}\sum_{\ell'} b_{i\ell'} A_{i\ell',n\ell}$$

$$\times \left(\frac{2\ell'+1}{2\ell+1}\right) e^{h(\nu_i-\nu_n)/kT_e} = b_{n\ell}\sum_{j=1 \text{ or } 2}^{n-1}\sum_{\ell''} A_{n\ell,j\ell''}, \tag{3.21}$$

which can be solved iteratively (Brocklehurst, 1970).

Substituting the solutions $n_{n\ell}$ into Eq. (3.2), we can obtain the intensity of the emission lines. The effective recombination coefficient for the recombination lines $n\ell \rightarrow n'\ell'$ is defined as

$$\alpha_{n\ell,n'\ell'}^{\text{eff}} = \frac{n_{n\ell} A_{n\ell,n'\ell'}}{n_p n_e}. \tag{3.22}$$

Note that $n_{n\ell}$ is a combination of α's and A's and is a function of T_e. The name "effective recombination coefficient" is unfortunate because it often gets confused with the recombination coefficient.

Although collisions are not important in the excitation of H, they are important in shifting between different angular momentum states of the same principal quantum number n. Such transitions will involve no energy change and can involve collisions with both protons and electrons. For example, at 10^4 K, the p-H collision cross section for $2S$ to $2P$ is $\sim 3 \times 10^{-10}$ cm^2. The additions of the collisional terms to Eq. (3.19) will provide a more accurate solution. Effective recombination coefficients taking into account collisions between different n (in particular $\Delta n \pm 1$) and L states are calculated by Hummer and Storey (1987; see Table 3.4).

For $T_e = 10^4$ K, $\alpha_{H\beta}^{\text{eff}}$ has values of 2.04 and 3.034×10^{-14} cm^3 s^{-1} respectively for Case A and Case B. The value is higher for Case B because the higher Lyman lines are recycled to Balmer photons.

Expressed in terms of the effective recombination coefficient, the integrated emission coefficient (j) is

$$4\pi j = 4\pi \int j_\nu d\nu$$

$$= n_e n_p \alpha^{\text{eff}} h\nu. \tag{3.23}$$

The emission coefficient ($4\pi j$) for Hβ in case B is $1.24 \times 10^{-25} n_e n_p$ erg cm^{-3} s^{-1}. From Table 3.4, the corresponding value for Brackett γ is $3.41 \times 10^{-27} n_e n_p$ erg cm^{-3} s^{-1}. Sometimes it is preferable to observe Brγ instead of Hβ because there is less extinction in the near-infrared part of the spectrum. An example of the H Brackett series recombination lines observed in PN is shown in Fig. 3.4.

The total flux in Hβ observed from an optically thin nebula of radius R at distance D is given by

$$F_{\text{H}\beta} = \frac{4\pi j_\nu}{4\pi D^2}\left(\frac{4\pi R^3\epsilon}{3}\right) \tag{3.24}$$

$$= \left(\frac{R^3\epsilon}{3D^2}\right) h\nu_{\text{H}\beta} n_e n_p \alpha_{\text{H}\beta}^{\text{eff}}, \tag{3.25}$$

Table 3.4. *Emission coefficients (relative to $j_{H\beta}$) for some of the lower transitions of H under Case B*

i/j	2	3	4	5	6	7	8
3	2.85						
4	1.00	0.332					
5	0.469	0.162	0.0777				
6	0.260	0.0901	0.0447	0.0245			
7	0.159	0.053	0.0275	0.0158	0.0093		
8	0.105	0.0365	0.0181	0.0104	0.00649	0.00401	
9	0.0734	0.0254	0.0126	0.00725	0.00456	0.00299	0.00191

For $T_e = 10^4$ K, $n_e = 10^4$ cm^{-3} (from Hummer and Storey, 1987).

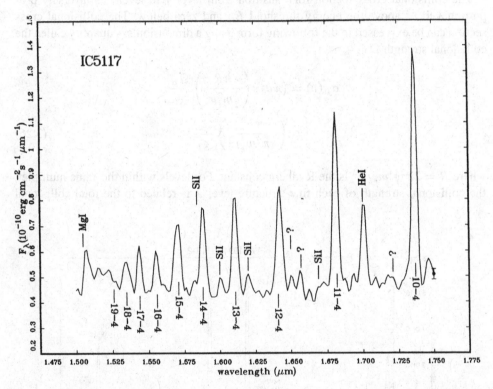

Fig. 3.4. The Brackett series of H recombination lines in IC 5117 (from Zhang and Kwok, 1992).

where the filling factor ϵ is introduced to take into account that the electron density may not be distributed uniformly over a sphere. For example, a shell of thickness $\Delta R = 0.2R$ can be approximated by setting $\epsilon = 0.6$. For $R = 0.1$ pc, $D = 1$ kpc, $\epsilon = 0.6$, $n_p = 10^3$ cm^{-3}, $n_e = 1.17n_p$, and $T_e = 10^4$ K, the total Hβ flux under Case B is 9×10^{-11} erg cm^{-2} s^{-1}.

3.5   Collisionally excited lines

Although the abundances of metals are much lower than that of hydrogen, collisionally excited atomic lines in PN are as strong as the H recombination lines because collisional processes are several orders of magnitudes faster than the recombination process. Collisional excitation to low-lying metastable states and their subsequence decay by means of magnetic dipole or electric quadrupole transitions lead to emission lines that are commonly referred to as forbidden lines.

Although PN are well known for their strong forbidden lines from ions such as N^+, O^+, O^{2+}, Ne^{2+}, Ne^{3+}, Ar^{2+}, Ar^{3+}, Ar^{4+}, and Mg^{4+}, forbidden lines are not the only collisionally excited lines in the PN spectrum. A number of near-UV permitted lines of metal (e.g. CII, CIV, NV, MgI, and MgII $3s\,^2S - 3p\,^2P$ 2,796/2,803 Å) are also collisionally excited and are very prominent in PN spectra observed by the *IUE* (Table 3.1). Some of these lines can be seen in the *IUE* spectrum of Hu1-2 shown in Fig. 3.5.

The collisional cross section for transition from level j to level i is inversely proportional to v^2 above the energy threshold E_{ij} and zero below. The collisional cross section can be expressed in the following form using a dimensionless quantity called the collisional strength Ω_{ji}:

$$\sigma_{ji}(v) = \left(\pi a_0^2\right) \left(\frac{hR}{\frac{1}{2}m_e^2 v^2}\right) \frac{\Omega_{ji}}{g_j},$$

$$= \left(\frac{\pi h^2}{4\pi^2 m_e^2 v^2}\right) \frac{\Omega_{ji}}{g_j}, \tag{3.26}$$

where $R = 2\pi^2 e^4 m_e / h^3$ is the Rydberg constant. For levels within the same multiplet, the collisional strength of each fine-structure level J is related to the total collisional

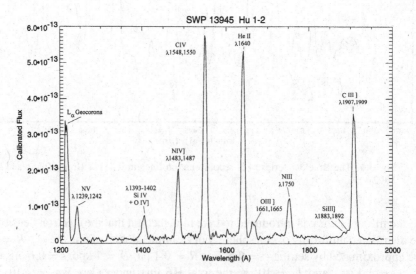

Fig. 3.5. The IUE short-wavelength prime spectrum of Hu1-2. The vertical scale is in units of erg cm^{-2} s^{-1} Å^{-1} (from W. Feibelman).

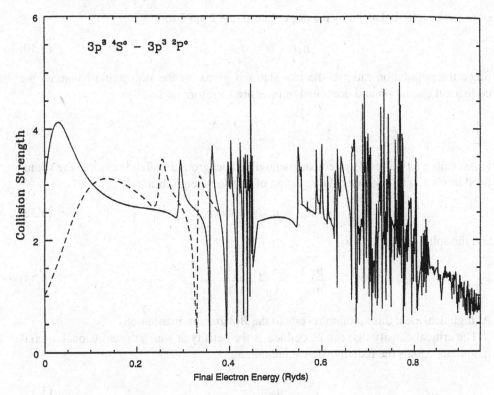

Fig. 3.6. Collisional strength for the $3s^2 3p^3\ ^4S - 3s^2 3p^3\ ^2P$ transition of SII. The solid and dashed curves are the calculations of Ramsbottom *et al.* (1996) and Cai and Pradhan (1993), respectively.

strength of the multiplet by

$$\Omega_{ji}(J) = \frac{(2J+1)}{(2L+1)(2S+1)}\Omega_{ji}. \qquad (3.27)$$

The collisional strengths consist of resonances that vary rapidly with energy (see Fig. 3.6). However, in astrophysical situations in which the colliding electrons have a range of energies, the effects of the resonances are averaged out. The collisional excitation rate for an ensemble of electrons at temperature T_e can be obtained by averaging over the Maxwellian distribution:

$$C_{ji}(T_e) = \int_{v_{min}}^{\infty} v\sigma_{ji}(v) f(v)\, dv, \qquad (3.28)$$

where $v_{min} = (2E_{ij}/m_e)^{1/2}$. Substituting Eq. (3.26) into Eq. (3.28), we have

$$C_{ji} = \frac{8.629 \times 10^{-6}}{T_e^{1/2}} \frac{\Omega_{ji}}{g_j} e^{-E_{ij}/kT_e}\ \text{cm}^3\ \text{s}^{-1}. \qquad (3.29)$$

At high densities, the population distribution is dominated by collisions and the principle of detailed balance requires that the excitation rate from a lower state 1 to an upper state 2 is exactly balanced by the deexcitation rate from 2 to 1:

$$n_2 n_e C_{21} = n_1 n_e C_{12}. \tag{3.30}$$

Since the population ratio of the two states is given by the Boltzmann equation, the collisional excitation and deexcitation rates are therefore related by

$$\frac{C_{12}}{C_{21}} = \frac{g_2}{g_1} e^{-E_{21}/kT_e}. \tag{3.31}$$

Assuming a two-level system and negligible background radiation (so that the stimulated terms can be omitted), the equation of statistical equilibrium is

$$n_2(A_{21} + n_e C_{21}) = n_1 n_e C_{12}, \tag{3.32}$$

and the solution is

$$\frac{n_2}{n_1} = \frac{n_e C_{12}}{A_{21}} \left[\frac{1}{1 + \frac{n_e C_{21}}{A_{21}}} \right]. \tag{3.33}$$

At high densities, this solution reverts to the Boltzmann distribution.

The critical density (n_c) can be defined as the density at which the collisional deexcitation rate equals the radiative de-excitation rate:

$$n_c = \frac{A_{21}}{C_{21}}. \tag{3.34}$$

At densities higher than n_c, collisional deexcitation becomes significant, and the forbidden lines will be relatively weaker as the density increases.

Because of the low metal abundances and the small transition rates, the collisionally excited lines are generally optically thin. The line emissivity is then given by Eq. (3.2). As an example, the [OII] doublet 3,726 Å / 3,728 Å can be considered as a pair of two-level atom systems because of the small energy difference of the upper states $^2D_{3/2}$ and $^2D_{5/2}$ (Fig. 3.2).

Assuming that $E_2 = E_3, \nu_{31} = \nu_{21}$ and no transition between level 3 and 2, the intensity ratio of the two lines is

$$\frac{I(3{,}726\,\text{Å})}{I(3{,}728\,\text{Å})} = \frac{n_3 A_{31}}{n_2 A_{21}},$$

$$= \frac{C_{13}}{C_{12}} \left(\frac{1 + \frac{n_e C_{21}}{A_{21}}}{1 + \frac{n_e C_{31}}{A_{31}}} \right). \tag{3.35}$$

At low densities ($n_e C_{21} \ll A_{21}$ and $n_e C_{31} \ll A_{31}$), we have

$$\frac{I_{31}}{I_{21}} = \frac{C_{13}}{C_{12}},$$

$$= \frac{\Omega_{1,3}}{\Omega_{1,2}}, \tag{3.36}$$

which implies that every collisional excitation is followed by a line emission. Using Eq. (3.27), we have

$$\frac{I_{31}}{I_{21}} = \frac{2\left(\frac{3}{2}\right)+1}{2\left(\frac{5}{2}\right)+1} = \frac{2}{3}. \tag{3.37}$$

At high densities, Eq. (3.35) gives

$$\begin{aligned}
\frac{I_{31}}{I_{21}} &= \frac{C_{13}}{C_{12}}\frac{C_{21}/A_{21}}{C_{31}/A_{31}}, \\
&= \frac{g_3 A_{31}}{g_2 A_{21}}, \\
&= \frac{2}{3}\frac{1.8\times10^{-4}}{3.6\times10^{-5}} = 3.3, \tag{3.38}
\end{aligned}$$

which is consistent with the expectation that the population distribution should obey the Boltzmann distribution, and the ratio of the populations of the two upper states is given by the ratio of their statistical weights.

Therefore, the very different behavior of the line ratios at different density regimes allows the [OII] doublet and other similar ions with outer electron configuration np^3 (e.g. [SII] λ 6,716 Å/6,731 Å) to be used as density probes of the nebula.

Another set of forbidden lines that are useful for nebular diagnostics are the $^1S_0 - {}^1D_2 - {}^3P_{0,1,2}$ transitions of ions with the np^2 electron configuration (e.g. [OIII] and [NII], see Fig. 3.7).

In the low-density limit, every collisional excitation will result in the emission of a photon, and the upper states can only be populated by collisional excitation from the

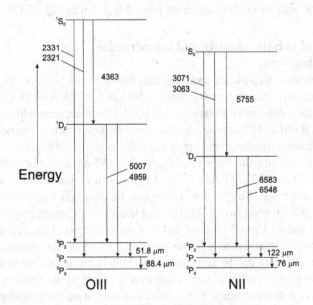

Fig. 3.7. The energy diagrams of [OIII] and [NII].

ground state. The equations of statistical equilibrium are

$$n_2 A_{21} = n_1 n_e C_{12}$$
$$n_3(A_{31} + A_{32}) = n_1 n_e C_{13}, \tag{3.39}$$

where the levels 1, 2, and 3 correspond to the P, D, S states, respectively. The solution to Eq. (3.39) is

$$\frac{n_3}{n_2} = \frac{C_{13}}{C_{12}} \frac{A_{21}}{A_{31} + A_{32}}. \tag{3.40}$$

The intensity ratio of the 4,363 Å to 5,007 Å + 4,959 Å lines is

$$\frac{I(4,363 \text{ Å})}{I(5,007 \text{ Å} + 4,959 \text{ Å})} = \frac{n_3 \nu_{32} A_{32}}{n_2 \nu_{21} A_{21}}. \tag{3.41}$$

Substituting Eqs. (3.29) and (3.40) into Eq. (3.41), we have

$$\frac{I(4,363 \text{ Å})}{I(5,007 \text{ Å} + 4,959 \text{ Å})} = \frac{\Omega_{13}}{\Omega_{12}} e^{-E_{32}/kT} \left(\frac{\nu_{32}}{\nu_{21}}\right) \frac{A_{32}}{A_{31} + A_{32}}. \tag{3.42}$$

With $E_{32} = 2.84$ eV, $A_{32} = 1.8$ s^{-1}, $A_{31} = 0.221$ s^{-1}, $\Omega_{13} = 0.28$, and $\Omega_{12} = 2.17$, we have

$$\frac{I(4,363 \text{ Å})}{I(5,007 \text{ Å} + 4,959 \text{ Å})} = 0.132 \times e^{-32,990/T_e} \tag{3.43}$$

The measurement of this line ratio therefore can serve as a probe to the electron temperature. At higher densities, collisional deexcitation terms have to be included in Eq. (3.39) and the line ratio will be slightly different from that given in Eq. (3.43).

3.6   Determination of nebular density and temperature by diagnostic diagrams

In the previous section, we saw the example that the ratio $I(4,363 \text{ Å})/I(4,959 \text{ Å} + 5,007 \text{ Å})$ gives the electron temperature for [OIII], whereas $I(3,726 \text{ Å})/I(3,729 \text{ Å})$ gives the electron density for a different ionization stage [OII]. However, monochromatic images of PN indicate that [OIII] and [OII] may originate from different parts of the nebula and the temperature and density measurements may not refer to the same region.

For a p^3 configuration as in [SII], the lines $I(6,717 \text{ Å})/I(6,730 \text{ Å})$, together with the transitions $I(4,068 \text{ Å} + 4,076 \text{ Å})/I(6,717 \text{ Å} + 6,730 \text{ Å})$ give both T_e and n_e in the same radiating volume for the same ion. Similar line ratios can be obtained from other p^3 ions such as [OII], [NeIV], [ClIII], and [ArIV]. With a combination of different transitions and ions, the values of T_e and n_e should be confined to a narrow region (see Fig. 3.8). The fact that the curves do not intersect at a single point suggests that the temperature and density distribution in PN may not be homogeneous. This is borne out by modern observations that the temperatures in PN in fact range over 6 orders of magnitude, and the densities also vary over a similar range from the dilute bubble to the low-density halo, to the high-density swept-up shell (see Chapter 13).

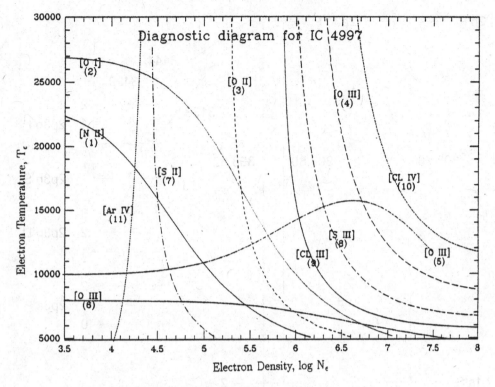

Fig. 3.8. Diagnostic diagram for IC 4997 (from Hyung *et al.*, 1994a).

3.7 Resonance fluorescence for OIII

Whereas the low-lying states of metals can be excited collisionally, the high excitation potential states of metals have low populations because of the slowness of the recombination process. In fact, permitted recombination lines from high-level states are generally weak except for H and He. Therefore, the observations of strong permitted lines from O^{++} in the near-UV were not expected. The excitation mechanism was identified by Bowen (1935) as being due to a wavelength coincidence between the $\mathrm{Ly}\alpha(\lambda\,303.780\,\text{Å})$ line of HeII and the $2p^2\ {}^3P_2 - 3d\ {}^3P_2^o$ $(\lambda\,303.799\,\text{Å})$ transition of OIII. Subsequent cascades to the $2s^2 2p3p$ $({}^3S_1$ and ${}^3D_{3,2,1})$ states and further cascades to the $2s^2 2p3s\ {}^3P_{2,1,0}$ give rise to the strong visible lines observed (Fig. 3.9).

Charge transfer (see Section 2.6) of O^{3+} in collision with H can populate the excited states of O^{2+}, in particular the $2p3p\ {}^3D$ and $2p3p\ {}^3S$ states that contribute to the Bowen cascade. The efficiency of the Bowen fluorescence process is measured by the fraction of all the HeII $\mathrm{Ly}\alpha$ photons created by recombination which are converted to photons in OIII transitions, excluding the $2p3p - 2p^2$ transitions. The typical observed efficiencies are ~ 0.1–0.3.

3.8 Forbidden lines of less-abundant elements

Since the identification of the forbidden lines of O by Bowen (1928), forbidden lines from ions of common elements ranging from N to Fe have been detected (Kaler *et al.*, 1976). Recent sensitivity improvements as the result of the use of CCD detectors have

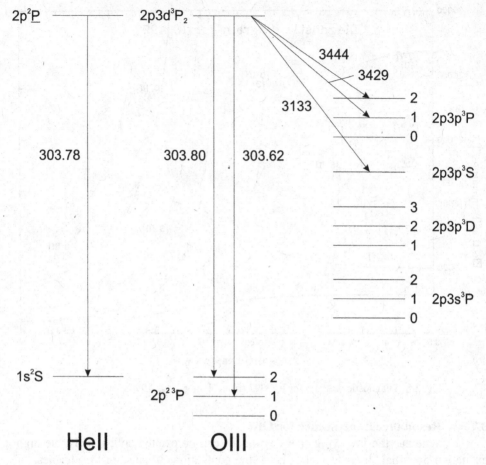

Fig. 3.9. Schematic partial energy diagram of OIII, illustrating the Bowen fluorescence. Not all levels or transitions are shown.

led to the discovery of lines of krypton (Kr, Z = 36), xenon (Xe, Z = 54), and other less-abundant elements in the fourth, fifth, and sixth row of the periodic table (Péquignot and Baluteau 1994). Probable identifications include selenium (Z = 34), bromine (Z = 35), rubidium (Z = 37), and barium (Z = 56). Assuming reasonable collisional strengths, Kr and Xe are estimated to be ~20 times more abundant in NGC 7027 than in the solar system.

3.9   Determination of the rest wavelengths of forbidden lines

The rest wavelengths of forbidden lines are difficult to determine accurately by laboratory measurements. Since the wavelengths of permitted lines such as Hα and HeI are known to high accuracies, the radial velocities of PN can also be determined accurately. Assuming that there is no systematic difference between the velocities of permitted and forbidden lines, the rest wavelengths of forbidden lines can be obtained to a high degree of precision by observing a large sample of PN. Table 3.5 lists the wavelengths offour forbidden lines derived from observations of over 60 nebulae.

Table 3.5. *Wavelengths in air at standard temperature and pressure for* [NII] *and* [SII]

Ion	Transition	Wavelength Å
NII	$^1D_2 - {}^3P_0$	6458.082 ± 0.008
	$^1D_2 - {}^3P_2$	6583.454 ± 0.006
SII	$^1D_{5/2} - {}^4S_{3/2}$	6716.472 ± 0.013
	$^1D_{5/2} - {}^4S_{3/2}$	6730.841 ± 0.013

Table taken from Dopita and Hua (1997).

3.10 Optical spectroscopic surveys of PN

PN can be recognized by their emission-line spectra and the absence of a continuum. Early surveys based on observations by objective prism were limited by sensitivity, and often only Hα could be detected in a faint nebula (Henize, 1967). Since Hα is not as reliable as forbidden lines as a indicator of PN, confusion with other objects often occurred. A survey of 51 PN in the Perek and Kohoutek catalog was carried out by Aller and Keyes (1987), who used an image-tube scanner on the 3-m telescope at Lick Observatory. From the measured line fluxes, they were able to derive electron temperatures, densities, and ionic abundances for these nebulae. A comprehensive survey of PN was carried out between 1984 and 1991 by Acker and Stenholm (cf. Acker *et al.*, 1992). The northern survey was carried out with the 1.93 m telescope of the Observatoire de Haute Provence and the southern survey at the 1.52 m telescope of the European Southern Observatory. An image dissector scanner and later a CCD camera were used in the survey. Over 1,000 PN were observed in the wavelength range 4,000–7,400 Å at a resolution of ∼10 Å. The excitation class, electron temperature, electron density, and the absolute Hβ fluxes were derived for a large number of PN (Acker *et al.*, 1989, 1991). The spectroscopic survey was also able to distinguish between true PN and other objects such as HII regions, symbiotic stars, and the like that were once misclassified as PN (Acker *et al.*, 1987).

4

Nebular continuum radiation

Several processes contribute to the continuum emission in PN. First, there is the stellar photospheric continuum. Because of the high temperatures of the central stars, the stellar continuum is strongest in the near UV. In the visible region, the main contributors are the free-bound (f-b) emissions from H and He in the nebular gas. In the infrared, free-free (f-f) emissions become stronger than the f-b emissions, and in the radio domain, the free-free process is the dominant emission mechanism in PN. Most interestingly, there is a continuum resulting from the two-photon decay of the 2^2S level of H, a process which is not observable in the laboratory environment.

4.1   Free-bound continuum radiation

The main contributor to the continuum radiation in the optical region in PN is f-b radiation. Every time an electron of mass m_e traveling at velocity v recombines with a proton to form an atom at level n, a photon of energy

$$h\nu = \frac{1}{2}m_e v^2 + \frac{h\nu_1}{n^2} \tag{4.1}$$

is emitted. The recombination rate to level n from electrons in the velocity range from v to $v + dv$ is

$$F_{\nu n}d\nu = n_p n_e f(v, T_e) v \sigma_n dv. \tag{4.2}$$

Making use of Eqs. (2.21), (4.1), and (2.29), we can obtain the emission coefficient

$$4\pi j_\nu = h\nu F_{\nu n}$$

$$= \frac{8\pi}{\sqrt{m_e}} \left(\frac{16}{3\sqrt{3}} \frac{e^2 h}{m_e^2 c^3}\right) (h^3 \nu_1^2) \left(\frac{1}{2\pi k}\right)^{3/2} \frac{n_p n_e}{T_e^{3/2}} \frac{g_{bf}}{n^3} e^{h(\nu_n - \nu)/kT}$$

$$= 2.16 \times 10^{-32} \frac{n_e n_p}{T_e^{3/2}} \frac{g_{bf}}{n^3} e^{h(\nu_n - \nu)/kT_e} \text{ erg cm}^{-3} \text{ s}^{-1} \text{ Hz}^{-1}. \tag{4.3}$$

The total emission coefficient, taking into account all levels, is

$$4\pi j_\nu = 2.16 \times 10^{-32} \frac{n_e n_p}{T_e^{3/2}} \sum_{n'}^{\infty} \frac{g_{bf}}{n^3} e^{h(\nu_n - \nu)/kT_e} \text{ erg cm}^{-3} \text{ s}^{-1} \text{ Hz}^{-1}, \tag{4.4}$$

where $\nu_{n'}$ must be less than ν. This is because any f-b photon that is due to recombination to level n' will have the minimum energy of $\nu_{n'}$. The emission coefficient for H at $T_e = 10^4$ K

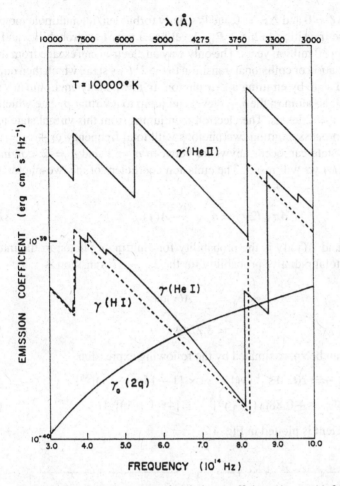

Fig. 4.1. Free-bound continuum emission coefficient $(4\pi n_e n_p j_v)$ for H, HeI, and HeII at 10,000 K (from Brown and Mathews, 1970).

is shown in Fig. 4.1. At λ 3,646 Å (Balmer limit), the f-b emission coefficient has a discontinuity known as the Balmer jump:

$$D_B = \log \frac{I^-}{I^+}, \tag{4.5}$$

where I^- and I^+ represent the continuum level at either side of v_2. The values of I^- and I^+ can be calculated by taking the ratio of j_v at $v = v_2$ for $n' = 2$ and 3 respectively. At $T_e = 10,000$ K, D_B has a value of 1.36. Observationally, D_B have been found to have values much less than 1.36 (Hua, 1974) which could be due to the rise in the continuum level by contribution from scattered light.

4.2 Two-photon radiation

The $^2S_{1/2}$ level of the hydrogen atom is metastable. Because of the spherically symmetric nature of the initial and final wave functions, radiative transition to the ground state 1^2S will have $\Delta\ell = 0$ and $\Delta m = 0$ and is strictly forbidden (all multipole moments are zero). A radiative transition to the $2^2P_{1/2}$ level has a very low probability and will occur only once every 30 million years. The only way an electron can escape from the $2s$ level is by photoionization or collisional transition to the $2^2P_{1/2}$ state, which then quickly decays to the ground state by emitting a Lyα photon. In low density environments such as those found in PN, an atom in the $^2S_{1/2}$ level can jump to a virtual p state which lies between $n = 1$ and $n = 2$ levels. The electron then jumps from this virtual state to the ground state, in the process emitting two photons with total frequency $\nu_1 + \nu_2 = \nu_{Ly\alpha}$. Since this virtual p state can occur anywhere between $n = 1$ and $n = 2$, continuum emission longward of Lyα will result. The emission coefficient of the two-photon (2γ) process is given by

$$4\pi j_\nu(2\gamma) = n_{2^2S}\frac{h\nu}{\nu_{Ly\alpha}}A(y), \tag{4.6}$$

where $y = \nu/\nu_{Ly\alpha}$ and $A(y)dy$ is the probability for emitting a photon in the interval $dy = d\nu/\nu_{Ly\alpha}$. The total radiative probability for the $2s \to 1s$ transition is

$$A_{2^2S,1^2S} = \int_0^1 A(y)dy$$

$$= 8.2249 \text{ s}^{-1}. \tag{4.7}$$

The function $A(y)$ can be approximated by the following expression

$$A(y) = 202.0 \text{ s}^{-1}\left(y(1-y)\times\{1-[4y(1-y)]^{0.8}\}\right.$$
$$\left. +0.88[y(1-y)]^{1.53}\times[4y(1-y)]^{0.8}\right). \tag{4.8}$$

The emission coefficient is plotted in Fig. 4.2.

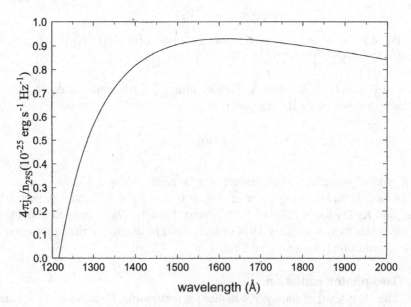

Fig. 4.2. Two-photon continuum emission coefficient ($4\pi j_\nu/n_{2^2S}$) for H.

Under Case B conditions, Lyα is trapped between the 2^2P and 1^2S states; every emission out of $2P$ is followed by an upward transition by stimulated absorption. The principal way out of the $2P$ state is through a collisional transition to 2^2S, followed by a 2γ emission. In this case, the population of the 2^2S state is maintained by a balance between recombination and 2γ radiation

$$n_p n_e (\alpha'_{2^2S} + \alpha'_{2^2P}) = n_{2^2S} A_{2^2S,1^2S}, \tag{4.9}$$

where α' is the total rate of populating a state by direct recombination or by recombinations to higher states followed by cascade to that state. Under Case A, the $2P$ state can depopulate by means of Lyα emission, and the $2S$ population is given by

$$n_p n_e \alpha'_{2^2S} = n_{2^2S} A_{2^2S,1^2S}. \tag{4.10}$$

At $T_e = 10^4$ K, α'_{2S} and α'_{2P} have values of 0.838 and $1.74\ 10^{-13}$ cm^3 s^{-1}, respectively. Since all cascades will end up in either 2^2S or 2^2P states, the sum of α'_{2S} and α'_{2P} must be equal to α_B (see Section 2.2). The level population n_{2^2S} obtained from these values using Eqs. (4.9) or (4.10) can then be substituted into Eq. (4.6) to calculate the spectrum of 2γ emission under Case A or Case B conditions.

A similar process also occurs in the singlet state of neutral and ionized He. The emission probability $A(y)$ for an hydrogenic ion with nuclear charge Z can also be approximated by Eq. (4.8) with the first constant replaced by $Z^6 \times R_Z/R_H \times 202$ s^{-1}. The total transition probability is

$$A_{2\gamma}^Z = 8.2249 Z^6 \frac{R_Z}{R_H}\ \text{s}^{-1}. \tag{4.11}$$

Figure 4.3 shows a simulated spectrum (near-UV to near-infrared) of an ionization-bounded PN. The photospheric continuum (assumed to be a blackbody) of the central star

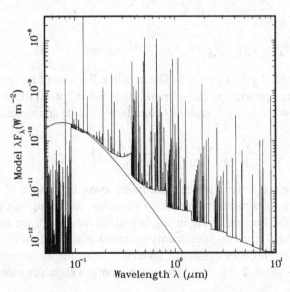

Fig. 4.3. The spectrum of an ionization-bounded PN with typical PN abundances calculated with the *CLOUDY* photoionization program. The blackbody spectrum of the central star is also plotted.

is also plotted. We can see that the stellar Lyman continuum is completely absorbed by the nebula and its fluxes are redistributed to the gas component in the form of emission lines and nebular continuum. The Balamer, Paschen, and so on jumps of the f-b continuum can clearly be seen, and the 2γ continuum can be seen in the near-UV. The emission spectrum beyond the Lyman limit is made up of nebular lines from higher ionization states of metals.

4.3   Free-free continuum emission

Whenever an electron passes near an ion, it will be accelerated and emits a radiation pulse. Since each free electron has a different energy and passes the ion at different distances, this results in a continuum radiation called thermal bremsstrahlung or free-free (f-f) radiation. The emission coefficient for f-f emission is given by

$$j_\nu = \int_\chi^\infty n_e(v)w(v)dv, \qquad (4.12)$$

where $n_e(v)$ is the velocity distribution of the electrons, $w(v)$ is the power radiated per electron with velocity v, and the lower limit of the integral

$$\chi = \left(\frac{2h\nu}{m_e}\right)^{1/2} \qquad (4.13)$$

is set by the fact that radiation at frequency ν can only arise from interactions in which the initial energy is greater than $h\nu$. Assuming a Maxwellian distribution for n_e, the emission coefficient at temperature T_e is then (Oster, 1961)

$$4\pi j_\nu = \frac{32\pi^{3/2}e^6 Z^2}{3^{3/2}m_e^2 c^3}\sqrt{\frac{2m_e}{kT_e}}g_{\rm ff}(\nu, T_e)n_i n_e e^{-h\nu/kT}$$

$$= 6.84 \times 10^{-38}Z^2 n_i n_e T_e^{-1/2}g_{\rm ff}(\nu, T_e)e^{-h\nu/kT} \text{ erg cm}^{-3}\text{ s}^{-1}\text{ Hz}^{-1} \qquad (4.14)$$

where Z is the nuclear charge and $g(\nu, T)$ is the f-f Gaunt factor. Because of the exponential term, f-f emission is more important at radio frequencies than at optical frequencies. At radio frequencies, $\exp(-h\nu/kT_e) \sim 1$, and the Gaunt factor can be approximated by

$$g_{\rm ff}(\nu, T_e) = \frac{\sqrt{3}}{\pi}\left\{17.7 + \ln\left[\frac{(T_e/K)^{3/2}}{(\nu/{\rm Hz})Z}\right]\right\}. \qquad (4.15)$$

Since the number densities of heavy elements are much lower than those of H and He, they only make minor contributions to the total f-f emission. Assuming that the He to H number ratio is y and the fraction of He in singly ionized form is y', we find that the singly ionized He to H ratio is $y'y$ and the doubly-ionized He to H ratio is $(1 - y')y$ if there is no neutral He. If the number density of ionized H is n_p, the electron density $n_e = x_e n_p$, where $x_e = 1 + yy' + 2y(1 - y')$. The mean atomic weight per electron is then

$$\mu_e = \frac{1 + 4y}{1 + yy' + 2y(1 - y')}. \qquad (4.16)$$

Taking into account the difference in Gaunt factors, we can modify the emission coefficient from that of pure H by the factor (Milne and Aller, 1975)

$$Y = 1 + \frac{n(\text{He}^+)}{n_p} + 3.7\frac{n(\text{He}^{++})}{n_p}. \tag{4.17}$$

For $y = 0.11$ and $y' = 0.5$, μ_e and Y have values of 1.236 and 1.258, respectively. The total emissivity at low frequency is therefore

$$4\pi j_\nu = 6.84 \times 10^{-38} n_e n_p g_{\text{ff}}(\nu, T_e, Z=1) T_e^{-1/2} Y \text{ erg cm}^{-3} \text{ s}^{-1} \text{ Hz}^{-1}. \tag{4.18}$$

Note that the emission coefficient has no frequency dependence other than in the Gaunt factor. The optically thin flux at distance D is

$$F_\nu = \frac{4\pi j_\nu}{4\pi D^2} \left(\frac{4\pi R^3 \epsilon}{3}\right). \tag{4.19}$$

For $R = 0.1$ pc, $D = 1$ kpc, $\epsilon = 0.6$, $n_p = 10^3$ cm^{-3}, and $T_e = 10^4$ K, the total flux received at 5 GHz is ~ 0.3 Jy, where Jy is a flux unit (10^{-23} erg cm^{-2} s^{-1} Hz^{-1}).

In the above expression for the total flux, we have assumed that the source function $B_\nu(T_e)$ is a constant throughout the nebula. The value of T_e is determined by the balance of heating by photoionization and cooling by forbidden-line and f-f radiation. Since the nebula is usually optically thick in the Lyman continuum (Section 2.1), photoionization is mostly due to diffuse Lyman continuum photons and not direct starlight. As a result, T_e is not a strong function of distance from the central star. Also the forbidden-line cooling rate increases with increasing T_e, which controls changes in T_e, confining the kinetic temperature to a limited range of values throughout the nebula.

Since the recombination lines and the f-f continuum radiation come from the same ionized region, the f-f flux is related to the flux of recombination lines such as Hβ. Taking the ratio of Eq. (3.25) and Eq. (4.19) and using the Hβ effective recombination coefficient for $T_e = 10^4$ K, we have

$$F_{\text{H}\beta}(\text{erg cm}^{-2} \text{ s}^{-1}) = 3.58 \times 10^{-10} Y^{-1} F_{5\text{GHz}}(\text{Jy}). \tag{4.20}$$

Under thermodynamical equilibrium, the absorption coefficient (κ_ν) is related to the emission coefficient by Kirchoff's law:

$$\kappa_\nu = \frac{j_\nu}{B_\nu}. \tag{4.21}$$

Combining Eqs. (4.14) and (4.21), we have for the absorption coefficient of H

$$\kappa_\nu = 3.69 \times 10^8 T_e^{-1/2} \nu^{-3} g(\nu, T_e, Z=1)\left(1 - e^{-\frac{h\nu}{kT_e}}\right) n_e n_p \text{ cm}^{-1}. \tag{4.22}$$

For $h\nu/kT_e \ll 1$,

$$\kappa_\nu = 0.0177 g(\nu, T_e)\nu^{-2} T_e^{-3/2} n_e n_p \text{ cm}^{-1}. \tag{4.23}$$

For a nebula with $R = 0.1$ pc and $n_{\text{H}} = 10^3$ cm^{-3}, the optical depth through the center of the nebula at 5 GHz is

$$\tau_\nu = 2R\kappa_\nu, \tag{4.24}$$

$$= 2.6 \times 10^{-3}. \tag{4.25}$$

Most PN are optically thin at 5 GHz, but some young nebulae are optically thick at this frequency. At the optically thick limit, the nebula resembles a blackbody at temperature T_e; therefore

$$F_\nu = \theta^2 \pi B_\nu(T_e) \tag{4.26}$$

$$= \pi \theta^2 \frac{2kT_e}{c^2} \nu^2 \quad \text{for} \quad \frac{h\nu}{kT_e} \ll 1, \tag{4.27}$$

where $\theta = R/D$ is the angular radius of the nebula. The slope of the radio spectrum (log F_ν vs. log ν) therefore changes from 2 at low frequency to $\sim$0 (in fact approximately -0.1 because of the Gaunt factor) at high frequency.

4.3.1   Total energy loss

The total energy loss rate caused by the f-f process over all frequencies is given by

$$\dot{E} = \int \int_0^\infty 4\pi j_\nu d\nu dV$$

$$= 6.8 \times 10^{-38} Y T_e^{-1/2} (n_e n_p \epsilon V) \int_0^\infty g_{ff} e^{-\frac{h\nu}{kT_e}} d\nu \text{ erg s}^{-1}$$

$$= 1.43 \times 10^{-27} T_e^{1/2} \bar{g}_{ff} Y n_e n_p \epsilon V \text{ erg s}^{-1}, \tag{4.28}$$

where $\bar{g}_{ff}$ is the average Gaunt factor and has a value of $\sim$1.3. In a nebula of radius 0.1 pc, filling factor 0.6, and H density 10^3 cm^{-3}, the total energy loss that is due to f-f is 1.2×10^{34} erg s^{-1}, or $\sim$5 $L_\odot$. Thus PN are very powerful radio sources.

4.3.2   Free-free radiation from a stellar wind

In a spherically-symmetric nebula with non-uniform density distribution, the observed flux at distance D is given by

$$F_\nu = \int_0^R \frac{B_\nu(T_e)}{D^2}[1 - e^{-\tau(p)}]2\pi p dp, \tag{4.29}$$

where R is the radius of the nebula and p is the impact parameter of the line of sight from the center of the nebula. If we write Eq. (4.23) as $\kappa_\nu = K_\nu n_e^2$, then

$$\tau(p) = K_\nu \int_{-\sqrt{R^2-p^2}}^{\sqrt{R^2-p^2}} n_e^2(r)dz, \tag{4.30}$$

where z is the length element along the line of sight. In a stellar wind, the density distribution is given by the equation of continuity

$$n_e(r) = \frac{\dot{M}}{4\pi \mu_e m_p r^2 V}, \tag{4.31}$$

where $\dot{M}$ is the mass-loss rate and V is the wind velocity. Assuming a constant $\dot{M}$ and V, then $n_e = Ar^{-2}$, Eq. (4.30) can be integrated ($R \to \infty$) to give

$$\tau(p) = \frac{\pi}{2} \frac{K_\nu A^2}{p^3}. \tag{4.32}$$

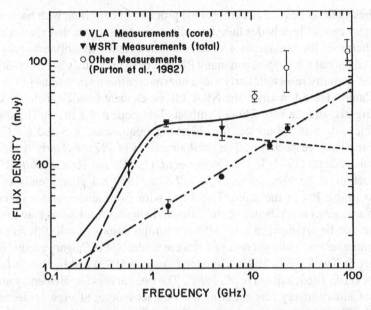

Fig. 4.4. The radio spectrum of M2-9. The three curves correspond to emissions from the core (dot-dash), the nebula (dash), and the total flux (solid) (from Kwok *et al.*, 1985).

Substituting Eq. (4.32) into Eq. (4.29), we have

$$F_\nu = 2\pi \left(\frac{\pi}{2}\right)^{2/3} \frac{A^{4/3}}{D^2} B_\nu(T_e) K_\nu^{2/3} \times \int_0^\infty \left(1 - e^{-\frac{1}{y^3}}\right) y \, dy \qquad (4.33)$$

The integral on the right has a numerical value of 1.33. Since $K_\nu \propto \nu^{-2}$, $B_\nu \propto \nu^2$ in the Rayleigh-Jeans limit, $F_\nu \propto \nu^{2/3}$. For $T_e = 10^4$ K,

$$F_\nu = 23 \left[\frac{(\dot{M}/M_\odot \, \mathrm{yr}^{-1})}{\mu_e(V/\mathrm{km \, s}^{-1})}\right]^{4/3} (\nu/\mathrm{Hz})^{2/3} \left(\frac{g_\nu}{x_e}\right)^{2/3} (D/\mathrm{kpc})^{-2} \mathrm{Jy}. \qquad (4.34)$$

The radio spectrum is flatter than that of a finite-size nebula because one sees into different depths of a stellar wind at different frequencies. Because the absorption coefficient is higher at lower frequencies, the size of the optically thick surface is larger at lower frequencies. This translates to a larger flux at low frequencies than in the finite-size nebula case.

Figure 4.4 shows the spectrum of the central-star wind from M2-9 fitted by Eq. (4.34).

4.4   Radio observations of planetary nebulae

Planetary nebulae have been extensively observed at radio wavelengths since the early days of radio astronomy (Terzian, 1966; Higgs, 1971). Systematic surveys of PN were carried out with the 64-m Parkes radio telescope at 2.7 GHz (Aller and Milne, 1972; Milne and Webster, 1979); 5 GHz (Milne and Aller, 1975; Milne, 1979), and at 14.7 GHz (Milne and Aller, 1982). Total fluxes for over 200 objects were obtained in these surveys.

Radio synthesis telescopes can circumvent the problem of confusion with background sources, and they can achieve better flux determinations for sources that are not overly resolved. Furthermore, the angular resolutions of synthesis telescopes often surpass those of ground-based optical telescopes, and many PN that have previously been classified as "stellar" can be spatially resolved. Early radio interferometric observations of PN were made at the Cambridge 5-km array, the NRAO three-element interferometer in Green Bank, West Virginia, and the Westerbork Synthesis Telescope in the 1970s. The smallest beam size achievable was ~2 arcsec, for observing frequencies of 5 and 8.1 GHz at Cambridge and NRAO respectively. The shell structures of PN are clearly revealed in the observations of Scott (1973, 1975), George *et al.* (1974), and Terzian *et al.* (1974).

The construction of the *Very Large Array (VLA)* in 1980 has greatly enhanced our capabilities to image PN in the radio. The *VLA*, with 27 antennas and a maximum baseline of 35 km, gives much better sensitivities and angular resolutions than previous radio arrays. It can be arranged in four different configurations to suit different target sizes. Two comprehensive radio surveys of PN were made, by the Calgary group (Kwok, 1985a; Aaquist & Kwok 1990, 1991; Kwok and Aaquist 1993) and by the Groningen group (Gathier *et al.*, 1983; Zijlstra *et al.*, 1989). The two surveys are divided by angular sizes, with the Calgary survey concentrating on compact sources of a few arcseconds or smaller, and the Groningen survey on sources from a few arcseconds to 1 arcmin. The Calgary survey was performed exclusively in the "A" (most extended) configuration, with angular resolutions of 0.4, 0.2, and 0.1 arcsec at λ 6, 3, and 2 cm respectively. The Groningen survey was obtained at λ 6 cm in the "B", "C" and "D" (most compact) configurations with angular resolutions of 1.2, 3.9, and 14 arcsec, respectively. Together, these two surveys have produced more accurate positions, fluxes, and angular sizes for several hundred PN. In addition, the Calgary survey has resolved many optically stellar PN and identified a population of very high surface brightness (and therefore likely to be young) PN. Examples of 15-GHz *VLA* maps of PN are shown in Fig. 4.5.

The *VLA* radio surveys of PN have largely been performed in snap-shot modes. This is generally sufficient for the three arms (and the 351 baselines) of the *VLA* to provide adequate $u - v$ coverages for small objects. However, the dynamic range is often limited to ~100 to 1, and higher dynamic range data can only be obtained by doing a full synthesis.

The absence of zero-baseline data also leads to an insensitivity to large-scale structures (e.g., halos) and the underestimation of total fluxes. In general, the *VLA* fluxes are superior to single-dish measurements because source confusion problems are avoided. However, in cases of extended and weak sources, it is easy to miss flux with interferometric measurements (Tylenda *et al.*, 1992, Pottasch and Zijlstra, 1994). Aaquist and Kwok (1990) used both summation of intensities from the map and the averaged visibility data extrapolated to the zero baseline to estimate the total fluxes. Although these two methods generally give consistent results within a few percent, the flux errors are larger for weaker, more extended sources. In these cases, the errors are estimated to be 10% in addition to the 10% systematic error in absolute flux calibration.

For the *VLA* in the 1980s, the λ 2 cm receiver sensitivity limit was ~0.1 mJy/beam, corresponding to a brightness temperature limit of ~20 K in the "A" configuration. In order to reveal large, low surface brightness structures, multifrequency and multiconfiguration observations are needed.

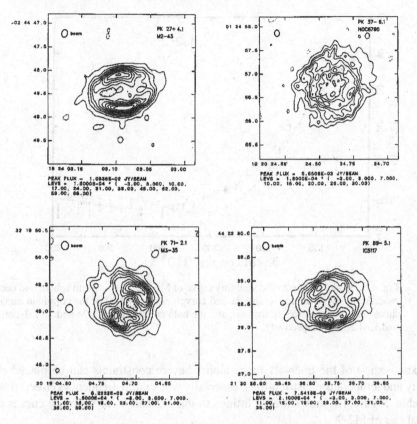

Fig. 4.5. Examples of high-resolution radio maps of compact PN (from Aaquist and Kwok, 1991).

4.4.1 *Visibility analysis of compact sources*

Although the majority of PN that have stellar appearances in the optical are resolved in the radio, a number of PN still remain unresolved. Some are only marginally resolved, and their structures cannot be easily determined from intensity maps. In some cases, the visibility curve analysis can be very useful. For a circular symmetric source, all the fluxes are in the real part of the visibility function. If this is the case, then the real component of the data can be averaged over angle and binned over projected baseline to improve the signal-to-noise ratio. The visibility $V(b)$ can be calculated for various intensity distributions [$I(r)$, where r is the angular radius of the source] by using the following equation:

$$ V(b) = 2\pi \int_0^\infty r I(r) J_0(2\pi r b) \mathrm{d}r, \qquad (4.35) $$

where b is the projected baseline and J_0 is the zeroth order Bessel function. The model visibility curve can then be compared to the observed visibility data. Figure 4.6 shows that the data for Me 2-2 can be better fitted by a core-halo model than by a uniform sphere.

Another useful application of the visibility curve analysis is for the identification of stellar winds. It is often difficult to distinguish between stellar winds and spheres by

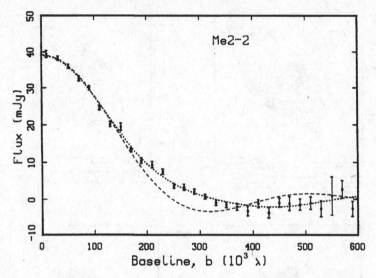

Fig. 4.6. Model fittings to the visibility curve of Me 2-2. The uniform sphere and core-halo models are shown in dotted and dashed curves, respectively. For the core-halo model, the fluxes and the diameters of the core and the halo components are 24 and 15 mJy, and 0.94 and 1.64 arcsec, respectively.

the examination of the intensity maps alone. Severe constraints can be placed on the density and temperature distributions, especially when multi-frequency observations are available. Figure 4.7 shows model fittings to the λ 2 and 6 cm visibility curves of the central star of M2-9.

4.5   Determination of the nebular mass

Whereas radio interferometric observations can image the structure of the ionized component of PN, radio photometric observations giving the total fluxes from f-f emissions have been frequently used to derive integrated properties of the nebula. For example, the ionized mass (M_i) of a spherically symmetric, uniform density PN is related to its proton density in the following way

$$M_i = \frac{4\pi}{3} n_p \mu m_H \epsilon R_i^3, \tag{4.36}$$

where R_i is the ionized radius of the nebula and $\mu = (1+4y)$ is the mean atomic weight per H atom. Substituting Eq. (3.25) into Eq. (4.36) we have

$$M_i = \frac{\mu m_H (4\pi D^2) F_{H\beta}}{h\nu_{H\beta} n_e \alpha_{H\beta}^{eff}}$$

$$= 11.7 \times 10^{11} (D/\text{kpc})^2 F_{H\beta}(\text{erg cm}^{-2}\,\text{s}^{-1})(n_e/\text{cm}^{-3})^{-1}\,M_\odot \tag{4.37}$$

assuming $T_e = 10^4$ K and $y = 0.11$. If the electron density can be determined by the forbidden-line ratios, then the ionized mass of the nebula can be estimated by measuring the extinction-corrected Hβ flux.

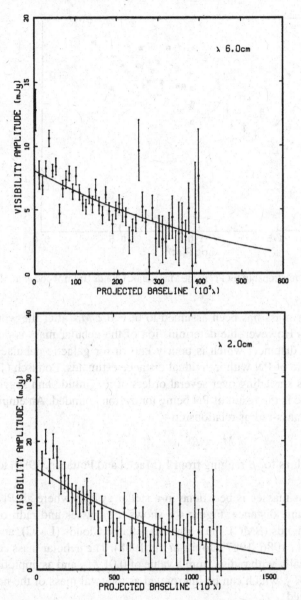

Fig. 4.7. λ 6 and 2-cm visibility curves of M2-9 fitted by a stellar wind model (Kwok *et al.*, 1985).

Alternatively, the ionized mass can be expressed in terms of the radio flux by combining Eqs. (4.18), (4.19), and (4.36):

$$M_i = \frac{\mu m_H (4\pi D^2) F_\nu}{6.84 \times 10^{-38} n_e g_{ff} T_e^{-0.5} Y}$$

$$= 282 \, (D/\text{kpc})^2 F_{5\,\text{GHz}}(\text{Jy})(n_e/\text{cm}^{-3})^{-1} \, M_\odot. \tag{4.38}$$

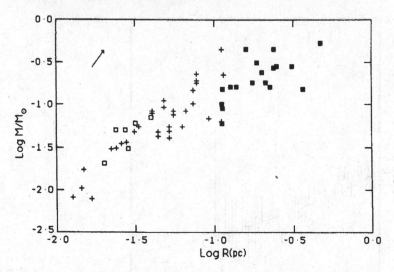

Fig. 4.8. Mass-radius relationship for PN in the Magellanic Clouds (from Wood *et al.*, 1987).

Traditionally the nebular mass has been assumed to be $\sim$0.2 $M_\odot$ after the work of Shklovsky (1956a, 1956b). However, the determination of the nebular mass is dependent on our knowledge of distance, which is poorly known for galactic nebulae (see Chapter 16). Using a sample of PN with individual distance estimates, Pottasch (1980) found that PN have masses spreading over several orders of magnitude and suggested that such a large mass range is the result of PN being ionization bounded. An empirical relationship known as the mass-radius relationship

$$M_i \propto R_i^\beta \qquad (4.39)$$

has been proposed, with values for β ranging from 1 (Maciel and Pottasch, 1980) to 5/3 (Daub, 1982).

A comparison of nebular masses is best done in a stellar system where all PN can be assumed to be at the same distance. Figure 4.8 shows the masses and radii of PN in the Small Magellanic Clouds (SMC), the Large Magellanic Clouds (LMC), and the galactic Center (Wood *et al.*, 1986, 1987; Dopita *et al.*, 1988). The nebular mass can be seen to increase monotonically with radius from a value of 0.01 $M_\odot$ and asymptotically approach a value of 0.27 $M_\odot$, which can be interpreted as the total mass of the nebula when it is completely ionized.

Another explanation for the mass-radius relationship is that the nebular mass in fact increases with age as the result of the interacting winds process (see Section 12.2). The fact that not all PN situated on the rising part of the mass-radius relationship are ionization bounded is emphasized by Méndez *et al.* (1992).

It is clear that the ionized nebular mass is not a static constant value but rather evolves with time as a result of photoionization and dynamics of the nebular expansion. It is also certain that both processes contribute to the increase of nebular mass with time, and the assumption that most PN are ionization bounded is too simplistic (see Section 17.1).

5

The neutral gas component

Since molecules were not expected to survive in the hostile high temperature, ionized environment of PN, the first detection of CO in the PN NGC 7027 by Mufson *et al.* (1975) therefore came as a complete surprise. The CO profile in NGC 7027 is almost 40 km s^{-1} wide, suggesting that the molecular gas is in an expanding envelope (Fig. 5.1). The amount of the molecular gas inferred from the CO line strength is over 1 $M_\odot$ and is much higher than the ionized mass ($\sim 0.1\ M_\odot$) usually associated with PN.

The CO profile of NGC 7027 resembles the CO profiles observed in the AGB stars, for example, IRC+10216 (Solomon *et al.*, 1971). By the mid-1970s, CO emission had been detected from many AGB stars (see Section 10.4.3). The amounts of mass observed in the circumstellar envelopes of AGB stars are found to be similar to that observed in NGC 7027. The most likely explanation of the origin of molecular gas in PN is that they are the remnants of the circumstellar envelopes of AGB stars (Kwok, 1982).

The presence of molecules implies that there is more material in PN than what is suggested by the optical images. Vibrational and rotational states of molecules can be excited either collisionally or radiatively, and the observations of these transitions allow new ways to probe the physical structure of PN. Because the vibrational and rotational states of molecules have much smaller energy separations than those of electronic states, the vibrational and rotational transitions occur respectively in the infrared and millimeter-wave regions of the electromagnetic spectrum instead of the visible and near-UV as in the case of electronic transitions. This in turn requires that new techniques be used in the observations.

5.1 Physics of diatomic molecules

The notation for the electronic structure of molecules is similar to that for atomic structure under LS coupling (Section 2.4). Each electronic state is designated by $^{2S+1}\Lambda_\Omega$, where S is the total electronic spin and Λ is the projection of the total electronic orbital angular momentum along the internuclear axis. The upper case Greek letters Σ, Π, Δ, and so on, are used to represent $|\Lambda| = 0, 1, 2, \ldots$, analogous to the notation used in atomic structure. Ω, the projection of the total electronic angular momentum onto the internuclear axis, is given by the sum of Λ and Σ, where Σ (not to be confused with the state designation for $\Lambda = 0$) is the projection of the electronic spin angular momentums along the internuclear axis. The value of Σ ranges from $-S$ to $+S$ in increments of 1. For example, a $^2\Pi$ state has $^2\Pi_{3/2}$ and $^2\Pi_{1/2}$ (corresponding to $|\Lambda| = 1$, $\Sigma = \pm 1/2$). Unlike atomic notation, the values of Ω, Λ, and Σ can be added algebraically rather than vectorially

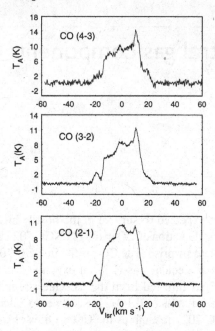

Fig. 5.1. The NGC 7027 CO $J = 4 - 3, 3 - 2$, and $2 - 1$ spectra obtained with the James-Clerk-Maxwell Telescope. The antenna temperatures have been corrected to outside of the Earth's atmosphere.

because they all refer to the same projection. If the electron wave function of a Σ state changes sign when reflected about any plane passing through both nuclei, it is denoted as Σ^-; if unchanged, then Σ^+. For a molecule with two identical nuclei, the right-subscript g (even) refers to the fact that the molecular wave function of this state remains unchanged on reflection through the center of the molecule, and u (odd) refers to a sign change.

The electronic states of diatomic molecules are also labeled with letters: X is used for the ground state, while A, B, C, and so on, are used for excited states of the same multiplicity $(2S + 1)$ as the ground state. States with a multiplicity different from that of the ground state are labeled with lower case letters $(a, b, c$, etc).

Because of the very different energies of the electronic, vibrational, and rotational states, these interactions can be assumed to be decoupled and their respective wave functions separable. The separation of nuclear and electronic wave functions is referred to as the Born-Oppenheimer approximation and its adoption greatly simplifies the handling of the quantum mechanics. Although this forms the basis of our following discussion, it should be remembered that effects ignored in making the Born-Oppenheimer approximation can yield interesting astrophysical results. The most well-known example is the interaction between the rotation of the nuclei and the electron orbital angular momentum, which leads to a splitting of the electronic state known as Λ doubling (see Section 5.6).

5.2   Rotational transitions

The energy levels of the rotational states of a diatomic molecule (such as CO) can be represented by the rigid body approximation,

$$E_J = hBJ(J + 1) \tag{5.1}$$

where J is the rotational quantum number and B is the rotational constant. For electric dipole transitions, only transitions between the successive rotational states are allowed ($\Delta J = \pm 1$). Therefore, the frequencies for the transition from upper state i to lower state j are given by

$$v_{ij} = 2B(J_j + 1). \tag{5.2}$$

For the CO molecule, B has the value of 57,563.56 MHz. The lowest rotational transition, $J = 1 \to 0$, has a frequency of 115 GHz. The spontaneous emission coefficient from the rotational state $J + 1$ to J for a diatomic molecule in the $^1\Sigma$ state is

$$A_{ij} = \frac{64\pi^4 v^3}{3hc^3} \mu^2 \frac{J_j + 1}{2J_j + 3}, \tag{5.3}$$

where μ is the permanent dipole moment of the molecule. For the $1 \to 0$ transition, A_{10} has the value of 7.4×10^{-8} s^{-1}, or 16 orders of magnitudes slower than the Lyα transition. In radio astronomy, it is common practice to define an excitation temperature (T_x) to represent the population ratio of two states

$$\frac{n_i}{n_j} = \frac{g_i}{g_j} e^{-hv_{ij}/kT_{x_{ij}}}, \tag{5.4}$$

where $g = 2J + 1$. If the rotational levels are thermalized (e.g., dominated by collisions with field particles), then T_x will be the same for all states and equal to the kinetic temperature of the field particles. However, in general the value of T_x is different for each transition. Substituting Eqs. (5.4) and (3.6) into Eq. (3.9), we have

$$a_v = \left(1 - e^{-\frac{hv}{kT_x}}\right) \frac{g_i}{g_j} \frac{c^2}{8\pi v^2} A_{ij}\phi_v$$

$$= \left(1 - e^{-\frac{hv}{kT_x}}\right) \frac{8\pi^3 v}{3hc} \mu^2 \frac{J_j + 1}{2J_j + 1} \phi_v. \tag{5.5}$$

If $T_x = 100$ K and ϕ_v is given by a square profile with a width of 1 km s^{-1}, then a_v has a value of 8.4×10^{-17} cm^2 for the $J = 1 \to 0$ transition of CO. Assuming $n_1 = 1$ cm^{-3}, the optical depth across a nebula of size 0.1 pc is 26.

The solution to the equation of transfer for a plane-parallel slab of material with a uniform source function is given by

$$I_v = I_{BG}e^{-\tau_v} + S_v(1 - e^{-\tau_v}), \tag{5.6}$$

where I_{BG} is the intensity of the background continuum radiation. Substituting Eq. (5.4) into Eq. (3.7), we can express the source function in terms of the excitation temperature as

$$S_v = \frac{2hv^3}{c^2} \frac{1}{e^{\frac{hv}{kT_x}} - 1}. \tag{5.7}$$

If we define the brightness temperature (T_b) as the temperature required for a blackbody to emit the same intensity at the same frequency, then Eq. (5.6) becomes

$$\frac{1}{e^{hv/kT_b} - 1} = \frac{e^{-\tau}}{e^{hv/kT_{BG}} - 1} + \frac{1 - e^{-\tau}}{e^{hv/kT_x} - 1}. \tag{5.8}$$

At low frequencies where the Rayleigh-Jeans approximation applies,

$$T_b - T_{BG} = (T_x - T_{BG})(1 - e^{-\tau}). \tag{5.9}$$

If the line is optically thick ($\tau \gg 1$), then the line temperature above the continuum is

$$T_b - T_{BG} = T_x - T_{BG}. \tag{5.10}$$

In this case, the excitation temperature is given by the measured brightness temperature of the line. If the molecule is thermalized at the kinetic temperature ($T_x = T_k$), then the kinetic temperature can be directly obtained. If the line is optically thin,

$$
\begin{aligned}
T_b - T_{BG} &= (T_x - T_{BG})\tau_\nu \\
&= \int (T_x - T_{BG}) n_j \phi_\nu \frac{8\pi^3\nu}{3hc} \mu^2 \frac{J+1}{2J+1} \left(1 - e^{-\frac{h\nu}{kT_x}}\right) dl
\end{aligned} \tag{5.11}
$$

The integrated line flux above the continuum is therefore proportional to the column density of the molecule in state j:

$$\int (T_b - T_{BG}) d\nu = (T_x - T_{BG}) \left(1 - e^{-\frac{h\nu}{kT_x}}\right) \mu^2 \frac{8\pi^3\nu}{3hc} \frac{J+1}{2J+1} \int n_j dl. \tag{5.12}$$

In the case of CO, the ^{12}CO line is generally optically thick. If the ^{13}CO line is optically thin, then the kinetic temperature obtained from the ^{12}CO measurement can be used in the ^{13}CO measurement to obtain the column density of the molecule and therefore the mass of the molecular gas in the nebula if the CO/H_2 abundance ratio is known.

Radio astronomers, carrying habits left over from the low frequency days, often use the term radiation temperature, which is defined as

$$T_R = \frac{c^2}{2\nu^2 k} I_\nu. \tag{5.13}$$

In that case

$$T_R = \frac{1}{e^{\frac{h\nu}{kT_x}} - 1} \frac{h\nu}{k}. \tag{5.14}$$

If the source is observed by a radio telescope with a beam (θ_A) larger than the source size (θ_s), the antenna temperature is given by

$$T_A = \left(\frac{\theta_s}{\theta_A}\right)^2 T_R, \tag{5.15}$$

if T_R and the beam profile are uniform over the source. If not, the antenna temperature is given by a product of the intensity distribution and the beam profile. For example, the main beam profile of a radio telescope can be represented by a Gaussian function

$$G(p) = \frac{4\ln 2}{\pi B^2} e^{-4\ln 2 p^2 / B^2}, \tag{5.16}$$

where p is the angular distance from the beam center and B is the half-power width of the beam. We should note that although T_R and T_A have units of temperature, they have the physical meaning of intensity.

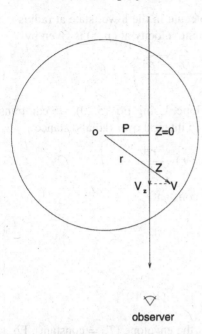

observer

Fig. 5.2. Schematic diagram of an expanding molecular envelope.

5.3 Molecular line profile in an expanding envelope

As the result of envelope expansion, the molecular lines observed in PN are often much wider than the thermal widths (ΔV) of the lines. Figure 5.2 shows a schematic diagram for an expanding envelope. If the expansion velocity is much greater than ΔV, then the molecules are not radiatively coupled with molecules in distant parts of the envelope. The intensity at each line-of-sight velocity (V_z) is then isolated to a path length $\Delta z(p)$, which corresponds to ΔV centered on V_z. The limits on Δz are given by z_1 and z_2.

The antenna temperature (T_A) at a given line-of-sight velocity (V_z) is the integral of the product of the telescope beam with the distribution of radiation temperature at that velocity. Assuming spherical symmetry of the envelope, the integration is most easily done over the impact parameter p:

$$T_A(V_z) = \int_0^{p_{max}} G(p) T_R(p, V_z) 2\pi p \, dp, \qquad (5.17)$$

where

$$T_R(p, V_z) = \frac{c^2}{2k\nu^2} [B_\nu(T_x) - B_\nu(T_{BG})] \left[1 - e^{-\tau(p, V_z)}\right], \qquad (5.18)$$

and p_{max} is the maximum value of p that can produce a line at velocity V_z. Here T_A refers to the antenna temperature above the off-source antenna temperature.

If we approximate the thermal line profile ϕ_ν by a square of width ΔV and height $1/\Delta V$, then the optical depth along the line of sight at velocity V_z is

$$\tau(p, V_z) = \int_{z_1}^{z_2} a_\nu n_j(r) dz, \qquad (5.19)$$

where $n_j(r)$ is the number density of the molecule in the lower state at radius r. For a velocity field $V(r) = V_0(r/R)^\beta$, the line-of-sight velocity at (p, z) is given by

$$V_z = V_0(r/R)^\beta \left(\frac{\sqrt{r^2 - p^2}}{r} \right), \tag{5.20}$$

where R is the radius of the molecular envelope. Using Eq. (5.20), we can transform Eq. (5.19) from a line-of-sight integration to an integral over radial distance

$$\tau(p, V_z) = \int_{r_1}^{r_2} a_\nu n_j(r) \frac{r\,dr}{\sqrt{r^2 - p^2}}. \tag{5.21}$$

For a uniformly expanding envelope ($V = $ constant)

$$V_z = \frac{V_z}{r}, \tag{5.22}$$

$$p = r\sqrt{1 - (V_z/V)^2}, \tag{5.23}$$

$$\Delta z = \frac{\Delta V}{dV_z/dz} = \Delta V \frac{p}{V}[1 - (V_z/V)^2]^{-3/2}. \tag{5.24}$$

Assuming a uniform excitation throughout the envelope ($T_x = $ constant), Eq. (5.17) can be integrated in the optically thick case ($\tau \to \infty$), yielding

$$T_A(V_z) = T_0\left\{1 - e^{-4\ln 2[1-(V_z/V)^2]R^2/B^2}\right\}, \tag{5.25}$$

where

$$T_0 = \frac{c^2}{2k\nu^2}[B_\nu(T_x) - B_\nu(T_{BG})]. \tag{5.26}$$

Figure 5.3 plots the profile of an optically thick line using Eq. (5.25) for the case in which the beam and source size are the same ($B = R$). At the center of the profile ($V_z = 0$), $T_A = T_0$. The excitation temperature can therefore be directly measured if the line is optically thick.

In a stellar wind of constant mass-loss rate ($\dot{M}$) and constant velocity,

$$n_j = \frac{g_j f \dot{M}}{4\pi V(2\,\mu m_H)r^2}$$

$$= \frac{A}{r^2}, \tag{5.27}$$

where μ is the mean atomic weight per H atom, and f is the molecule (e.g., ^{13}CO) to H_2 abundance ratio and g_j is the fraction of the molecule in the jth rotational state.

Substituting Eq. (5.27) into Eq. (5.21) and integrating, we have

$$\tau(p, V_z) = \frac{a_\nu A}{p}\left[\cos^{-1}(p/r)\right]_{r_1}^{r_2}$$

$$= \frac{a_\nu A}{p}\left[\cos^{-1}\sqrt{1 - (V_z/V)^2}\right]_{V_z - \Delta V/2}^{V_z + \Delta V/2}. \tag{5.28}$$

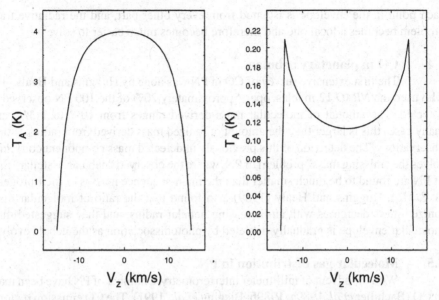

Fig. 5.3. Theoretical profile for molecular emission in a uniformly expanding envelope. Left, optically thick case; right, optically thin case.

For an optically thin line (e.g., ^{13}CO), $\tau \ll 1$, Eqs. (5.18) and (5.26) give

$$T_R(p, V_z) = T_0 \tau(p, V_z). \tag{5.29}$$

Substituting Eqs. (5.28) and (5.29) into Eq. (5.17), we have

$$T_A(V_z) = \frac{4\sqrt{\ln 2}}{B} a_v T_0 A \left[\cos^{-1} \sqrt{1 - (V_z/V)^2} \right]_{V_z-\Delta V/2}^{V_z+\Delta V/2} \times \int_0^{x_{max}} e^{-x^2} dx, \tag{5.30}$$

where

$$x = \frac{2p}{B} \sqrt{\ln 2} \tag{5.31}$$

and

$$x_{max} = \frac{2R\sqrt{\ln 2}}{B} \sqrt{1 - (V_z/V)^2}. \tag{5.32}$$

Figure 5.3 shows the profile of an optically thin line for $B = R$. If T_x is known from an optically thick line, then the mass-loss rate can be determined from the antenna temperature of an optically thin line.

Realistically, the molecules are excited by a combination of collisional excitation and radiative excitation (e.g., by absorption of dust continuum photons), and the excitation temperature is unlikely to be uniform throughout the envelope. In this case, a full radiation transfer calculation is needed. A useful approximation is the Sobolev approximation, in which the Doppler shifts introduced by the velocity field far exceed the local Doppler width created by thermal and microturbulent motions. Under the Sobolev approximation, each point in the envelope is isolated from every other part, and the radiative transfer problem becomes a local one and therefore becomes much easier to solve.

5.4   CO in planetary nebulae

The first extensive survey of CO in PN was done by Huggins and Healy (1989), who used the *NRAO* 12 m telescope. Approximately 20% of the 100 PN observed were detected. The amount of molecular mass derived ranges from 10^{-3} to 1 $M_{\odot}$, and in many cases this is larger than the amount of ionized mass derived from radio continuum observations. The detection of this previously undetected mass component conveniently solves the "missing mass" problem in PN, where the observed "nebular + stellar" masses of PN are found to be much smaller than the main-sequence masses of their progenitors (1–8 $M_{\odot}$). Huggins and Healy (1989) also found that the ratio of molecular mass to ionized mass decreases with an increasing nebular radius, and they suggested that the molecular envelope is gradually depleted by photodissociation as the nebula evolves.

5.5   Molecular gas distribution in PN

With the advent of millimeter interferometry, a number of PN have been mapped in CO (Bachiller *et al.* 1989a, 1989b; Bieging *et al.*, 1991). The CO emission regions are generally found to lie outside of the ionized nebula, confirming that the molecular gas represents the remnant of the AGB envelope. The ionized nebula of NGC 7027 ($\sim 10''$ in size) fits neatly into the central minimum of the CO distribution (Bieging *et al.*, 1991). A similar molecular cavity corresponding to the size of the ionized region has also been seen in IRAS 21282 + 5050 (Likkel *et al.*, 1994; Deguchi, 1995). The CO map of M1-7 (Shibata *et al.*, 1994) shows a double-peaked structure, very similar to the radio continuum structure of most PN. An inner bright ring is seen in the CO map of NGC 7027, suggesting that the molecular gas is compressed by the shock front created by the interacting winds process (Bieging *et al.*, 1991). A comparison of the CO and H_2 maps of NGC 6720 also suggests that a shock is propagating into the molecular envelope (Greenhouse *et al.*, 1988).

5.6   OH in planetary nebulae

The OH molecule has an unpaired electron and therefore possesses a net spin angular momentum ($S = 1/2$). The ground state for the OH molecule is $^2\Pi_{3/2,1/2}$, with the $\Omega = 3/2$ state having a lower energy than the $\Omega = 1/2$ state. The total angular momentum quantum number J, including rotation, will be 3/2, 5/2, 7/2, and so on for $\Omega = 3/2$, and 1/2, 3/2, 5/2, and so on for $\Omega = 1/2$. The interaction of the electron angular momentum and the rotational angular momentum splits each of these levels into two states, called Λ doubling. The interaction between the spin of the unpaired electron with the magnetic moments of the nuclei further splits each Λ doublet into two hyperfine states ($F = J + I$, where I is nuclear spin and has the value of 1/2). For the ground state $^2\Pi_{3/2}$, this results in four transitions between the two Λ doublets for $J = 3/2$ ($\Delta F = 0, \pm 1$) at 1,612.231 ($F = 1 \to 2$), 1,665.401 ($F = 1 \to 1$), 1,667.358 ($F = 2 \to 2$), and 1,720.528 MHz ($F = 2 \to 1$). Because of the small energy separations between these states, population inversion can occur as the result of infrared pumping to an excited vibrational state of OH. This leads to maser action, making the intensities of the lines much stronger and therefore much easier to detect. These maser lines, in particular the 1,612 MHz line, are commonly detected in O-rich AGB stars (see Section 10.4.2).

OH emission from PN was first detected in Vy2-2 (Davis *et al.*, 1979). The observed profile is similar to the double-peaked profiles observed in AGB stars except that the red-shifted component is missing. This asymmetry could be due to absorption of the receding component by the ionized gas, or preferential amplification of the continuum by stimulated emission for the approaching component (Kwok, 1981b).

Since OH emission is the result of maser action and relies on the existence of a long, velocity-coherent path to operate, it is not obvious how long after the end of the AGB phase that the OH line will remain detectable. The calculations of Sun and Kwok (1987) and Bujarrabal *et al.* (1988) show that the OH line will remain saturated in the first thousand years of post-AGB evolution. This is confirmed by the detection of OH emission in a number of young PN by Zijlstra *et al.* (1989).

5.7 Molecular hydrogen emission

Since the H_2 molecule has two identical nuclei, nuclear spin plays a role in the spectra of H_2. The two nuclear spins can be either parallel ($I = 1$, called the ortho state), or antiparallel ($I = 0$, called the para state). Since each of the nuclei is a fermion ($I = 1/2$), the total wave function (ψ_{tot}) must be antisymmetric upon exchange of the two nuclei. Under the Born-Oppenheimer approximation, the total wave function is a product of the nuclear, electronic, vibrational and rotation wave functions:

$$\psi_{tot} = \psi_{nuc}\psi_{el}\psi_{vib}\psi_{rot}. \tag{5.33}$$

Since ψ_{vib} is symmetric because the nuclear separation is unchanged on simple exchange of nuclei, and ψ_{el} is symmetric for the ground state of H_2, ψ_{rot} must be antisymmetric for ortho hydrogen and symmetric for para hydrogen. Therefore, ortho H_2 has only rotational states of odd J, and para H_2 has only rotational states of even J. In statistical equilibrium, the population ratio between ortho and para states is the ratio of $2I + 1$, or 3 to 1. The line intensity, which is proportional to the statistical weight, is $3(2J + 1)$ for odd J, and $(2J + 1)$ for even J.

The vibrational-rotational energy levels of a diatomic molecule under the rigid rotator and harmonic oscillator approximations are given by

$$E_{v,J} = (v + 1/2)h\nu_0 + J(J + 1)hB, \tag{5.34}$$

where ν_0 is the oscillator frequency. The selection rule for vibrational transitions within the harmonic approximation is $\Delta v = \pm 1$. The transitions are organized into branches according to the change in the rotational quantum number (initial minus final in an emission process) ΔJ. The selection rule for one-photon, electric-dipole transitions are $\Delta J = -1$ and $+1$ (called P and R branches, respectively). The frequencies of the transitions in the fundamental mode are as follows

$$\nu = \nu_0 + 2(J + 1)B, \qquad J = 0, 1, 2, \ldots \text{ for the } R \text{ branch,}$$

$$\nu = \nu_0 - 2JB, \qquad J = 1, 2, 3, \ldots \text{ for the } P \text{ branch.}$$

The hydrogen molecule has two identical nuclei and has no permanent electric or magnetic dipole moment. The only observable transitions are electric quadrupole transitions, which have $\Delta J = -2, 0$, and 2 (called O, Q, and S branches, respectively). Some examples of the vibrational-rotational transitions of H_2 are shown in Fig. 5.4.

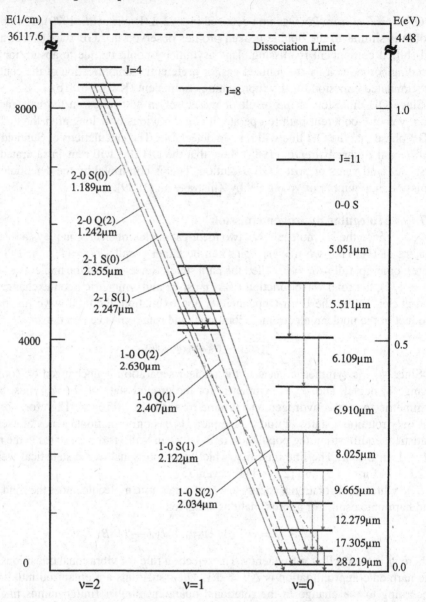

Fig. 5.4. The energy diagram of molecular hydrogen with some of the observed transitions marked (wavelengths in μm).

In the interstellar medium, the H_2 molecule is generally in the ground electronic state $X\ ^1\Sigma_g^+$. The selection rule for electronic dipole transitions of a diatomic molecule is $\Delta\Lambda = 0, \pm1$, $\Delta S = 0$, $\Delta\Sigma = 0$, and $\Delta\Omega = 0, \pm1$. Furthermore, only $g \leftrightarrow u$, $\Sigma^+ \rightarrow \Sigma^+$ and $\Sigma^- \rightarrow \Sigma^-$ are allowed. In the interstellar medium H_2 is generally dissociated by radiative dissociation. The $\Delta S = 0$ rule prevents the transition to the first unbound state $b^3\Sigma_u^+$, thereby protecting the molecule from photodissociation. Electronic transitions to the bound excited states $B\ ^1\Sigma_u^+$ and $C\ ^1\Pi_u$ (called the Lyman and Werner

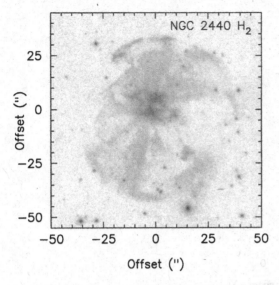

Fig. 5.5. Narrow-band H_2 image of NGC 2440 (from Latter and Hora 1997).

bands, respectively), and the subsequent radiative decays to high vibrational states of the ground electronic state, can result in dissociation. Most of the time, however, the molecule falls back to a bound ($v < 15$) vibrational state and emits a series of infrared photons as it cascades through the bound states in a process known as fluorescent radiation (Black and van Dishoeck, 1987). The population distribution of H_2 in various v and J states and line strengths of the fluorescent lines can be calculated by using a procedure similar to that used for the recombination lines of the H atom (see Section 3.4).

In comparison with collisional excitation, radiative excitation by the absorption of UV photons leads to the large population of the upper vibrational levels, resulting in relatively high fluxes in the lines with $\lambda < 2\,\mu$m. The different predicted line strengths between thermal collisional excitation and nonthermal radiative excitation can allow us to distinguish between these two mechanisms by infrared spectroscopy.

The 2.121 μm $v = 1 - 0$, $J = 3 - 1$ $S(1)$ electric quadruple vibration-rotation line of H_2 is widely observed in PN (Latter *et al.*, 1995). It is excited either by collisional excitation in shocked gas or by fluorescent emission (Dinerstein *et al.*, 1988). The H_2 line is therefore useful as a probe of the dynamics or the radiation distribution in PN. It has been suggested that PN of "butterfly" or "bipolar" morphology have a higher probability of having H_2 emission (Kastner *et al.*, 1996). This correlation could be explained if PN of these morphological types (see Chapter 8) have higher progenitor masses, which have higher mass-loss rates and shorter transition times between the AGB and PN phases (see Chapter 11), hence an inherently large mass of H_2, very little of which has been dissociated.

The development of infrared arrays has made possible the imaging of molecular hydrogen in PN. Figure 5.5 shows the narrow-band H_2 image of NGC 2440. The H_2 emission can be seen in a nearly circular ring with radial spokelike features. The location of H_2 emission in relation to the ionized region is best illustrated in NGC 7027 (Fig. 5.6). Two shells of H_2 emission can be seen: the inner one surrounds the HII region while the

NGC 7027

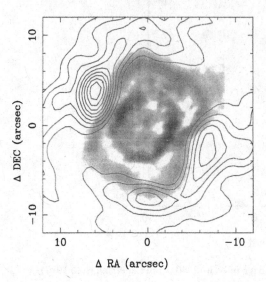

Fig. 5.6. Contours of CO ($J = 1 - 0$) emission plotted on the H_2 emission grey-scale map of NGC 7027 (from Graham *et al.*, 1993).

outer one is coincident with the inner edge of the CO envelope. The H_2 emission region actually represents the transition region from the ionized shell to the molecular shell.

The detection of molecular gas in PN suggests that, in addition to the ionization boundary that is due to the HII to HI transition, there also exists a photodissociation region (PDR) where H_2 is decomposed into H atoms by interactions with UV photons (Hollenbach and Tielens, 1997). The dissociation energy of H_2 is 4.48 eV, so photons between 912 and 2,770 Å can escape from the HII region and dissociate H_2. The PDR therefore lies between the ionized shell and the molecular envelope, in which H_2 is dissociated into HI. CO, having a dissociation energy of 11.09 eV, will remain in molecular form even after H_2 is dissociated. This transition is illustrated in the CO and H_2 maps of NGC 7027 (Fig. 5.6).

5.8   Emission from neutral atoms

Early studies of neutral material in the interstellar medium were based on several strong optical resonance lines (e.g., CaII H and K lines at λ 3,933 and 3,966 Å; and NaI D lines at λ 5,889 and 5,895 Å). With the advent of radio astronomy, the HI line at λ 21 cm became the natural choice. However, the search for HI in PN was difficult because of beam dilution and contamination by interstellar gas. It was not until the 1980s that HI (Schneider *et al.*, 1987) and NaI (Dinerstein and Sneden, 1988) lines were detected. Since the spectrum is usually dominated by strong interstellar lines, atomic lines associated with the near side of an expanding PN shell can be identified by searching for weak features blueward of the PN systemic velocity. Most of the strong resonance lines of abundant species lie in the UV. Absorption lines (e.g., CI and OI) have been detected in BD $+30°$ 3039 by the *IUE* (Pwa *et al.*, 1986) and the *Hubble Space Telescope* (*HST*, Dinerstein *et al.*, 1994).

Neutral atoms (CI, OI) can also be detected by their fine-structure lines. As a result of the small energy separations, these lines generally lie in the infrared. Examples are $^3P_1 - {}^3P_2(63 \ \mu m)$ and $^3P_0 - {}^3P_1(145 \ \mu m)$ lines of OI, and the $^3P_1 - {}^3P_0(609 \ \mu m)$ line of CI. Since CI is a signature of the PDR, it is expected in PN with thick molecular envelopes. CI was first detected in NGC 7027 (Young *et al.*, 1993) and is also found in NGC 6720 (Bachiller *et al.*, 1994) and NGC 7293 (Young *et al.*, 1997).

5.9 Circumstellar chemistry

The detection of CO and other molecules (e.g., NH_3, HCN, HCO^+, and CO^+) in PN suggests that molecules can survive for a long time after the initiation of photoionization. The molecules are protected from the UV radiation by dust absorption, and also by self-shielding and mutual shielding (for H_2 and CO). Early chemical models (Bachiller *et al.*, 1988; Howe *et al.*, 1992) predict that CO will be rapidly destroyed (in $< 10^2$ yr) by stellar UV radiation once the PN phase begins. However, to treat the chemistry in PN properly, we need to have a self-consistent model including photoionization, photodissociation, and chemistry in both the dust and gas phases. In particular, if the dynamical aspects are considered (see Chapter 13), the swept-up gas shell can be ionization bounded, and the UV photon density in the remnant AGB wind may be very low. This will allow the molecules to survive for 10^3 yr or more if the mass-loss rate of the AGB wind is high enough. Furthermore, the higher density and temperature conditions created by the shocked (but neutral) AGB wind will allow gas-phase reactions to take place For example, C^+ can react with H_2, forming CH^+, which through subsequent reaction with H_2 gives rise to the series of molecular ions CH_2^+, CH_3^+. These undergo recombination with electrons, as well as photodissociation, giving the neutrals CH and CH_2. The molecules CH^+, CH_2^+, CH_3^+, CH, CH_2 are important in leading to the formation of CN and HCN, two of the molecules observed in young PN (Bachiller *et al.*, 1997), through reactions with atomic nitrogen released from N_2:

$$CH^+ + N \rightarrow CN^+ + H$$
$$CN^+ + H \rightarrow CN + H^+$$
$$CH_2^+ + N \rightarrow HCN^+ + H$$
$$HCN^+ + H \rightarrow HCN + H^+$$
$$HCN^+ + e^- \rightarrow CN + H$$
$$CH + N \rightarrow CN + H$$
$$CH_2 + N \rightarrow HCN + H. \tag{5.35}$$

It is not clear whether detected CN and HCN in young PN are mostly remnants from their circumstellar chemistry or later re-formed in the photon dominated neutral envelopes of PN.

In addition to shock chemistry, interesting chemistry can also occur in the photodissociation region. In an ionization-bounded PN, the ionization front lags behind the photodissociation front of H_2, which heats the gas to $\sim 1,000$ K through photoelectric heating (dust $+ h\nu \rightarrow$ dust$^+ + e^-$) and H_2 photodissociation heating. At these temperatures, atomic oxygen released from CO reacts with remaining molecular hydrogen to form OH:

$$O + H_2 \rightarrow OH + H. \tag{5.36}$$

Once OH is formed, the following reactions lead to re-formation of CO and significant levels of CO^+ and HCO^+:

$$C^+ + OH \rightarrow CO^+ + H$$
$$CO^+ + H \rightarrow CO + H^+$$
$$CO^+ + H_2 \rightarrow HCO^+ + H$$
$$HCO^+ + e^- \rightarrow CO + H. \tag{5.37}$$

It now appears that the major formation process for CO^+ is the above sequence rather than the simple ionization of CO (which has an ionization energy of 14.1 eV). The above sequence explains strong HCO^+ emission observed in NGC 7027, NGC 6720, NGC 6781, and NGC 7283. The signature of photodissociation chemistry can be witnessed in the detection of HCO^+ in AFGL 618, but the same molecule is absent in AFGL 2688, an object in an earlier phase of evolution (see Chapter 14).

6

The dust component

The infrared spectrum of PN was expected to be dominated by forbidden-line emissions from the ionized gas, and the discovery of strong infrared excess in NGC 7027 was totally unpredicted (Gillett *et al.*, 1967). A photometric survey by Cohen and Barlow (1974, 1980) using the 1.5-m telescope at Mt. Lemmon showed that many PN display strong infrared emission from dust. Far infrared photometry observations from the *Kuiper Airborne Observatory (KAO)* by Telesco and Harper (1977) found the dust in NGC 7027 to be cool, with a color temperature of ∼100 K. Further *KAO* observations by Moseley (1980) confirmed the presence of cool dust in 13 PN.

The discovery of cool dust in PN and the observations of circumstellar dust envelopes in AGB stars (Section 10.4.1) suggest that they share the same origin (Kwok, 1982). If PN descend from mass-losing AGB stars, then the remnants of the AGB dust envelope must still be present in PN. The dispersal of the dust envelope since the end of the AGB implies a gradual decrease of the dust temperature, and the shifting of the peak of the dust continuum from mid-infrared to the far-infrared. According to the Wien's law,

$$\lambda_{\max} = \frac{2,900}{(T/K)} \mu m, \tag{6.1}$$

a blackbody of 100 K will peak at 30 μm, beyond the longest infrared window observable from the ground. Since the flux decreases exponentially on the short wavelength side of the Planck function, dust emission from PN is difficult to detect from the ground. The wide presence of dust in PN was only confirmed after the launch of the *IRAS* mission.

6.1   Dust absorption and emission

If the radius (a) of a dust particle is much greater than the wavelength of the photons it is interacting with, then the absorption coefficient is simply given by its geometric cross section

$$\kappa_\nu = \pi a^2 n_d, \tag{6.2}$$

where n_d is the number density of the dust particles. When the $a \gg \lambda$ condition is not valid, a dimensionless number $Q_\nu(a)$ can be introduced into Eq. (6.2) to give

$$\kappa_\nu = \pi a^2 Q_\nu(a) n_d. \tag{6.3}$$

The value for Q can be calculated by Mie theory (van de Hulst, 1957) and can be approximated by $0.1(\lambda/1\,\mu m)^{-\alpha}$, where α has the value of ∼1–2, depending on the composition of the grain. Assuming that the specific gravity of the grain is ρ_s, ψ is the dust to gas mass ratio, and the dust is uniformly distributed over the entire spherical volume V, then

$$n_d = \frac{\psi M_s}{\left(\frac{4\pi}{3}a^3\rho_s\right)V}, \tag{6.4}$$

where M_s is the mass of the nebular mass. The optical depth through the center of the nebula is

$$\tau = \int_{-R}^{R} \pi a^2 Q_\nu n_d dl. \tag{6.5}$$

For a nebula mass (M_s) of 0.2 $M_\odot$, radius (R) 0.1 pc, $\alpha = 1$, $a = 1\,\mu m$, $\rho_s = 1\,g\,cm^{-3}$, and $\psi = 10^{-3}$, τ has a value of 1.5×10^{-4} at the wavelength of 10 μm. PN are generally optically thin in the infrared. For comparison, the optical depth due to f-f at 10 μm is much smaller [see Eq. (4.23)]. Dust absorption is therefore the dominant absorption mechanism in the infrared.

In thermodynamic equilibrium, the emission coefficient is related to the absorption coefficient by Kirchoff's law

$$j_\nu = (\pi a^2 Q_\nu)n_d B_\nu(T_d), \tag{6.6}$$

where T_d is the dust temperature. Assuming optically thin conditions, the flux emitted from a nebula of radius R at distance D is

$$F_\nu = \frac{\psi M_s Q_\nu B_\nu(T_d)}{\frac{16}{3}\pi a \rho_s D^2}. \tag{6.7}$$

For $T_d = 150$ K and $D = 1$ kpc, the flux at 25 μm is 5.6 Jy.

Dust in PN is heated by a combination of direct starlight, Lyα photons, and nebular continuum photons. Since these fluxes vary with distance from the star, the dust temperature is not expected to be uniform throughout the nebula. For dust grains at distance r from the star, the heating rate (Γ) is

$$\Gamma = \int_0^\infty \kappa_\nu (4\pi J_\nu)d\nu, \tag{6.8}$$

where J_ν is the mean intensity at the location of the grains. Assuming that direct starlight is the dominant heating source,

$$J_\nu = \frac{1}{4\pi}\int B_\nu(T_*)d\Omega$$

$$= \frac{1}{4\pi}\int_0^{2\pi} d\phi \int_0^{R_*/r} B_\nu(T_*)\sin\theta d\theta$$

$$= \frac{B_\nu(T_*)}{2}[-\cos\theta]_0^{R_*/r}. \tag{6.9}$$

At great distances from the star, $R_*/r \ll 1$,

$$J_\nu = \frac{1}{4}\left(\frac{R_*}{r}\right)^2 B_\nu(T_*). \tag{6.10}$$

The cooling rate (Λ) by dust self-emission is

$$\Lambda = \int_0^\infty 4\pi j_\nu d\nu \tag{6.11}$$

Substituting Eqs. (6.3) and (6.6) into Eqs. (6.8) and (6.11) and balancing heating and cooling, we have

$$\int_0^\infty \frac{1}{4}\left(\frac{R_*}{r}\right)^2 (Q_\nu \pi a^2) B_\nu(T_*) d\nu = \int_0^\infty (Q_\nu \pi a^2) B(T_d) d\nu. \tag{6.12}$$

Since most of the photons emitted by the central star have wavelengths much smaller than the grain size, $Q \sim 1$ for the absorption and can be taken out of the integral on the left side, giving

$$\frac{1}{4}\left(\frac{R_*}{r}\right)^2 \left(\frac{\sigma T_*^4}{\pi}\right) = \int_0^\infty Q_\nu B_\nu(T_d) d\nu, \tag{6.13}$$

where σ is the Stephan-Boltzmann constant. If Q_ν is expressed as a power law ($Q_\nu \propto \nu^\alpha$), then the right-hand side of Eq. (6.13) can be integrated, giving the following relationship between the dust temperature and distance from the star:

$$T_d \propto r^{-\frac{2}{4+\alpha}}. \tag{6.14}$$

If we take into consideration the variation of dust temperature as a function of radius, then the emergent spectrum of PN would represent a composite of modified blackbodies each radiating at temperature T_d at radius r. The observed flux would then be

$$F_\nu = \frac{\int 4\pi j_\nu dV}{4\pi D^2}$$

$$= \frac{\pi a^2 Q_\nu}{D^2} \int_{R_*}^\infty n_d(r) B_\nu[T_d(r)] 4\pi r^2 dr. \tag{6.15}$$

For $n_d \propto r^{-n}$ and $\alpha = 1$, Eq. (6.15) gives $F_\nu \propto \nu^{2.5n-3.5}$ over the central part of the flux distribution. For $n = 2$, which corresponds to an equal amount of dust at each T_d, $\nu F_\nu \propto \nu^{2.5}$.

Although the dust component is generally optically thin in the infrared, it can be optically thick in the visible wavelengths. If this is the case, then the dust at the outer radii will see a decreasing amount of direct starlight, and it will be heated more and more by infrared photons emitted by nearby grains. In that case, the emergent flux has to be calculated by a complete self-consistent radiation transfer model (Leung, 1976).

The possibility that dust in PN can be heated by Lyα photons was first suggested by Krishna Swamy and O'Dell (1968). Since the optical depth of the Lyα line is very high ($>10^6$; see Section 3.4), a Lyα photon is scattered many times in the nebula and therefore increases its chance of being absorbed by dust. The luminosity of the Lyα line is

$$L(Ly\alpha) = 4\pi j_{Ly\alpha} V$$

$$= V h\nu_{Ly\alpha} n_e n_p \left(\sum_{n=2}^\infty \alpha_n - \alpha_{2^2S}^{\text{eff}}\right). \tag{6.16}$$

The sum begins at $n = 2$ for the first term because recombination to the ground state will not produce a Lyα photon. At high density ($n_e > 10^4$ cm^{-3}), the second term is dropped because atoms in the $2^2 S$ state can make collisional transition to the $2^2 P$ state and therefore create a Lyα photon (see Section 4.2). At $T_e = 10^4$ K, the first term has a value of 2.60×10^{-13} cm^3 s^{-1} and the second term has a value of 8.38×10^{-14} cm^3 s^{-1}. The emission coefficients ($4\pi j$) for Lyα are therefore 4.3 and 2.9×10^{-24} erg cm^{-3} s^{-1} in the high- and low-density limits, respectively.

Taking into account the contribution by Lyα photons, we can write Eq. (6.12) as

$$\pi \left(\frac{R_*}{r}\right)^2 \int_0^{\nu_1} (Q_\nu \pi a^2) B_\nu (T_*) d\nu + \frac{L_{Ly\alpha}}{n_d V} = 4\pi \int_0^\infty (Q_\nu \pi a^2) B(T_d) d\nu. \quad (6.17)$$

The upper limit in the first integral now goes only to ν_1 because all Lyman continuum photons are assumed to be absorbed in the gas component in an ionization bounded nebula. The infrared excess (*IRE*; Pottasch, 1984) can be defined as the ratio of L_{IR}/L(Lyα), where L_{IR} is the total luminosity radiated by the dust component. Since the Lyα line is difficult to observe, *IRE* can be expressed in terms of the f-f flux. Dividing $4\pi j_{Ly\alpha}$ by Eq. (4.18) and substituting into the definition of IRE, we have

$$IRE = A \frac{F_{IR}(10^{-11} \text{ erg cm}^{-2} \text{ s}^{-1})}{F_{5GHz}(\text{mJy})} \quad (6.18)$$

where the coefficient A has the value of 1.0 and 1.5 at high and low densities, respectively.

6.2   Dust continuum emission from planetary nebulae

Since the radiation from the dust component usually peaks beyond 20 μm, a comprehensive study of the dust in PN is best done from space. From the *IRAS* photometry of 46 nebulae, Pottasch *et al.* (1984) found that dust in PN has a wide range of temperatures (from 40 to 240 K), with the compact (and presumably younger) nebulae having higher temperatures. Using the radio fluxes (see Section 4.4) to extrapolate the expected level of nebular emission, Kwok *et al.* (1986) found that the near-infrared emissions from young nebulae are well fitted by nebular emission alone and are easily separated from the dust emission. Figure 6.1 shows the spectral energy distribution (SED) of the young PN IC 2448, showing the relative contributions of the nebular and dust components to the total emission.

The SED of 66 compact nebulae were fitted by a model consisting of stellar, nebular and dust components by Zhang and Kwok (1991). They found that on the average $38 \pm 21\%$, $25 \pm 16\%$, and $37 \pm 14\%$ of the emergent fluxes are emitted by the stellar, nebular, and dust components, respectively. All but nine of the 66 PN have IRE values greater than 1, suggesting that direct starlight must play a role in the heating of the dust.

6.3   Dust features

A family of strong unidentified infrared (UIR) emission features at 3.3, 6.2, 7.7, 8.6, and 11.3 μm was first detected in the young carbon-rich PN NGC 7027 (Russell *et al.*, 1977). These features are now identified as vibrational modes of various functional groups (Table 6.1). For example, these features are prominent in the infrared spectra of polycyclic aromatic hydrocarbon (PAH) molecules (Allamandola *et al.*, 1989).

Table 6.1. *Identification of the UIR features*

ν (cm^{-1})	λ (μm)	FWHM (cm^{-1})	Mode
3040	3.29	30	Aromatic $=$C–H stretch ($v = 1 \rightarrow v = 0$)
1615	6.2	30	Aromatic C$=$C stretch
1315–1250	7.6–8.0	70–200	Bending of several strong aromatic C-C stretching bands
1150	8.7	...	Aromatic $=$C–H in-plane bend
890	11.2	30	Aromatic $=$C–H out-of-plane bend for nonadjacent, peripheral H atoms

Table taken from Allamandola *et al.* (1989).

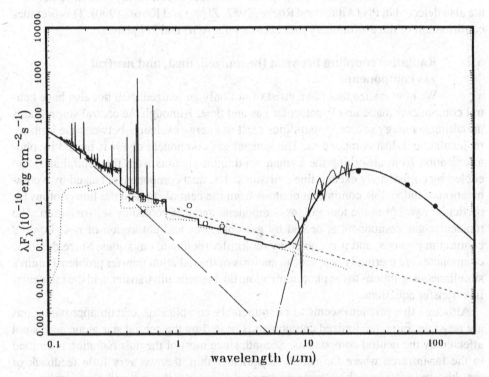

wavelength (μm)

Fig. 6.1. The SED of IC 2448. The dotted curve shows the model nebular (b-f and f-f) continuum, the dash curve is the f-f continuum level extrapolated from radio measurements, and the dot-dash curve is the central star. The total model curve is shown as a solid curve (from Zhang and Kwok, 1991).

PAH molecules are benzene rings linked to each other in a plane, with H atoms or other radicals saturating the outer bonds of peripheral C atoms. Absorption of UV photons results in a rapid redistribution of the photon energy among the vibrational modes of the molecule. In a low-density environment where collisional deexcitation is not possible, the molecules undergo spontaneous deexcitation by means of infrared fluorescence, leading to the infrared features observed. Since these features are not seen in AGB stars, the substances bearing the relevant functional groups (e.g. PAH molecules) must either be synthesized during the transition from the AGB to the PN phase, or they must be produced in the AGB atmosphere but only excited in the PN environment.

It is well documented that the 3.3-μm feature observed in the L band window is closely correlated with the presence of the 11.3-μm feature observed by the *IRAS* Low Resolution Spectrometer (LRS; Jourdain de Muizon *et al.*, 1989). The 6.2-μm and 7.7-μm UIR features have also been observed in PN from airborne observations (Cohen *et al.*, 1989). The UIR features are particularly strong in PN with large infrared excesses, and those with [WC 11] (see Section 7.2.1) central stars (Kwok *et al.*, 1993). The [WC] nature of the central star suggests that these are carbon-rich objects, and the large infrared excesses suggest these PN must be young and have low dynamical ages and are likely to have evolved from AGB stars with large core masses.

The 9.7-μm silicate and the 11.3-μm SiC feature, both commonly seen in AGB stars, are also detected in PN (Aitken and Roche, 1982; Zhang and Kwok, 1990). This provides confirmation of the evolutionary link between the AGB and PN phases.

6.4   Radiative coupling between the ionized, dust, and neutral gas components

We now realize that PN consist of not only an ionized shell but also have neutral components made up of molecular gas and dust. Although the central star remains the ultimate energy source, possibilities exist of energy exchange between the ionized, molecular, and dust components. The ionized gas component of PN is heated by photoionization from absorbing the Lyman continuum photons from the central star, and cooled by collisionally excited line emissions. The dust component is heated by a combination of the visible continuum photons from the central star and the line photons (in particular Lyα) from the ionized gas component, and it is cooled by self-radiation. The molecular gas component is heated by gas collisions and absorption of near-infrared continuum photons, and it is cooled by molecular rotational emissions. Since the three components are thermally coupled, an analysis of the radiation transfer problem requires simultaneous solutions to the photoionization, dust continuum transfer, and the molecular line transfer equations.

Although this problem seems to be intractably complicated, certain approximations are possible. First, the ionized component is heated by the central star alone and is not affected by the neutral components. Second, since most of the dust radiation is emitted in the far-infrared where the nebula is optically thin, there is very little feedback of the dust emission into the other components. Similarly, the molecular gas emits only line radiation; although these lines can be optically thick because of self-absorption, they do not constitute a significant source of heating of the dust component. However, near-infrared emission from the dust can be absorbed by the molecular gas by means of vibrational excitations, so the molecular line transfer problem is dependent on the output of the dust transfer problem.

A complete self-consistent model taking into account the radiative interactions of the ionized, dust, and neutral gas components was first made by Volk and Kwok (1997). Three different numerical codes for photoionization, dust continuum transfer, and molecular line transfer are integrated to calculate the overall spectrum between λ 1,000 Å to 1 cm

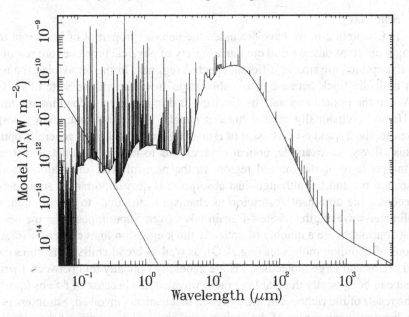

Fig. 6.2. The model spectrum of NGC 7027 from UV to radio. the thin (almost straight) line on the left is the continuum (226,000 K blackbody) of the central star.

of NGC 7027. First, the ionized gas component is treated independently. All the permitted lines with significant optical depths (e.g., the recombination lines of H and He) have their line transfer problem solved with the escape probability method. The output of the photoionization model, including the fluxes of the recombination lines of H and He, the collisionally excited lines of metals, b-f and f-f continua, and the stellar continuum, is fed to the dust continuum transfer problem as the input radiation field. The dust continuum transfer equations are then solved, producing continuum averaged intensities (J_ν) at every radius. The molecular gas is assumed to be mixed with the dust component, and the values of $J_\nu(r)$ are used as a source of background radiation available for stimulated absorption by the molecules. The equation of transfer and the equation of statistical equilibrium are then solved for the relevant molecules (e.g., CO) to produce the emergent fluxes in each of the rotational transitions.

Figure 6.2 shows the model spectrum of NGC 7027. Most of the lines in the UV, visible, and the infrared are recombination lines of H and He and collisionally-excited lines of metals. In the submillimeter region, the rotational lines of CO and H_2O are plotted. In the mid- and far-infrared, the continuum emission is dominated by the dust component. The major UIR features at 3.3, 7.7, 8.6, and 11.3 μm can be seen as narrow features. At wavelengths longer than λ 400 μm, the f-f continuum begins to be stronger than the dust continuum. The visible and near-infrared continua are marked by the H and He b-f jumps. The broad feature near 2,400 Å is the 2γ continuum. The absorption feature seen in the near-UV is the 2,200 Å feature. The blackbody spectrum of the central star is plotted as a dotted line. This figure clearly illustrates the richness of PN spectrum and why NGC 7027 is one of the most interesting celestial objects to observe.

6.5   Summary

In Chapters 2-6, we have discussed the nebular properties of PN. From multi-wavelength observations, we find out that a variety of physical processes are responsible for the absorption/emission in different spectral regimes. In the ultraviolet, the nebula is often optically thick because of b-f absorption from the ground state of H. In the near-UV and the optical regions, the spectrum is dominated by recombination lines of H and He and collisionally excited lines of metals. The major continuum absorption processes are the 2γ and b-f processes of H and He, and the nebula is generally optically thin. This allows, for example, optical observations to map out the detailed structure of the morphology of the ionized region. In the near-infrared, the major absorption processes are b-f and f-f, although dust absorption is also important in some nebulae. Dust becomes the dominant absorption mechanism in the mid- to far- infrared, up to the millimeter-wavelengths where f-f again takes over. Superimposed on the optically thin dust continuum are a number of optically thick emission lines caused by rotational transitions of common molecules (e.g., CO), as well as broad emission features caused by solid grains and large molecules. PN are generally optically thin between 1 μm and 1 cm, but can be optically thick at long radio wavelengths because of f-f absorption.

As the result of the richness of the emission mechanisms involved, astronomers have at their disposal many means of determining the physical properties of the nebula. For example, the emission measure of the ionized gas can be determined by measurements of the recombination lines and the f-f continuum fluxes. The electron density and temperature can be determined by measuring the ratios of certain forbidden lines. The dust temperature can be obtained by infrared photometry. The size and structure of the ionized and molecular gas can be obtained by optical/radio imaging and millimeter-wave imaging, respectively.

All these techniques also reveal a more complicated system than previously envisaged. The temperature of different regions of PN range from $\sim 10^1$ K in the molecular gas, $\sim 10^2$ K in the dust component, $\sim 10^4$ K in the ionized gas, $\sim 10^5$ K in the atmosphere of the central star, and $> 10^6$ K in the shocked stellar wind bubble (see Section 12.2). Correspondingly, all states of matter from solid-state, molecular, atomic, and ionic forms are present. The original photospheric continuum radiation from the central star interacts with all these different forms of matter and is reprocessed into emissions covering the entire electromagnetic spectrum, from the X-ray to radio, making PN some of the most interesting objects to study observationally.

7

Observations of the central star of planetary nebulae

Central stars of PN (CSPN) are difficult to study because of their faintness in the visible (due to their high temperature) and the contamination of their spectra by nebular emissions. Unlike stars on the main sequence (MS), for which there exists a unique relationship between mass and effective temperature, CSPN undergo considerable changes in temperature over their short lifetimes. Although their masses do not change significantly, their surface abundances do change as the result of nuclear burning and mass loss. Assuming that the stellar winds from the CSPN are driven by radiation pressure, the mass-loss rate is mainly a function of T_* and surface gravity (log g). In this case, the spectral classification of central stars can in principle be determined by three parameters: effective temperature, surface abundance, and wind strength. In this chapter, we discuss these three parameters in turn.

7.1   Determination of the temperature of the central star

7.1.1   Zanstra temperature

Zanstra (1927) developed the method to derive the central star temperature by comparing the nebular recombination flux with the stellar continuum magnitude. This method is based on the assumption that the number of Lyman continuum photons absorbed in the nebula is equal to the total number of recombinations to all levels excluding the ground state. Harman and Seaton (1966) used the nebular Hβ flux to estimate the total number of Lyman continuum photons, and they derived T_* by comparing the Hβ flux with the stellar V magnitude.

The ratio of the total nebular Hβ flux to the stellar continuum flux at V band is

$$\frac{F(\text{H}\beta)}{F_\nu} = \frac{h\nu_{\text{H}\beta} \int n_e n_p \alpha_{\text{H}\beta}^{\text{eff}} dV}{4\pi R_*^2 \pi B_\nu(T_*)}. \tag{7.1}$$

Substituting Eqs. (2.36), (2.41), and (2.42) into Eq. (7.1), we have

$$\frac{F(\text{H}\beta)}{F_\nu} = \frac{15h\nu_{\text{H}\beta}}{\pi^5 k B_\nu(T_*)} \sigma \frac{\alpha_{\text{H}\beta}^{\text{eff}}}{\alpha_B} T_*^3 G(T_*). \tag{7.2}$$

Since the ratio of $\alpha_{\text{H}\beta}^{\text{eff}}$ to α_B is independent of T_e, the central star temperature T_* can be determined from Eq. (7.2) by iteration.

The Zanstra method has been applied to more than 300 nebulae (Kaler, 1983; Shaw and Kaler, 1989; Gleizes *et al.*, 1989; Gathier and Pottasch, 1989). Since the Zanstra method assumes that the nebula is optically thick in the Lyman continuum, and since PN change from optically thick to optically thin in H and He at different times, this can lead to different estimates of the central star temperature by using H or He lines. The fact that stellar atmospheres are not well approximated by blackbodies can also contribute to errors in the Zanstra temperatures (Henry and Shipman, 1986, see Section 7.4). However, it can be argued that since CSPN have a large variety of spectral types and abundances, the blackbody assumption is probably no worse than any specific set of atmosphere models (Kaler, 1983). The main problem remains, however, that this method requires the measurement of the stellar magnitude. Since the nebular continuum is often brighter than the stellar photospheric continuum, a measurement of the stellar V magnitude can have very large errors, or be simply impossible.

7.1.2   The Stoy temperature

The energy-balance method was first introduced by Stoy (1933). It avoids the assumption of optically thick HI (or HeII) as required by the Zanstra method, and instead it uses the nebular forbidden- and recombination-line ratios to determine the central-star temperature. Assuming that the mean energy of the electron is $\frac{3}{2}kT_e$, then threshold energy available for the excitation of forbidden lines is

$$h\nu_e = h\nu_1 + \frac{3}{2}kT_e. \tag{7.3}$$

Since each photoionization can produce $h(\nu - \nu_e)$ amount of energy available for the excitation of forbidden lines, the total amount of energy emitted in forbidden lines is therefore

$$4\pi D^2 F(\mathrm{FL}) = 4\pi R^2 \int_{\nu_1}^{\infty} \frac{\pi B_\nu(T_*)}{h\nu} h(\nu - \nu_e)\,d\nu, \tag{7.4}$$

where $F(\mathrm{FL})$ is the total flux observed in forbidden lines. The total power radiated in the recombination line Hβ is

$$4\pi D^2 F(\mathrm{H}\beta) = h\nu_{\mathrm{H}\beta} \int n_e n_p \alpha_{\mathrm{H}\beta}^{\mathrm{eff}}\,dV. \tag{7.5}$$

Dividing Eq. (7.4) by Eq. (7.5) and making use of Eqs. (2.35) and (2.36), we have

$$\frac{F(\mathrm{FL})}{F(\mathrm{H}\beta)} = \frac{\alpha_B}{\alpha_{\mathrm{H}\beta}^{\mathrm{eff}} h\nu_{\mathrm{H}\beta}} \frac{\int_{\nu_1}^{\infty} B_\nu(T_*)\left(\frac{\nu - \nu_e}{\nu}\right) d\nu}{\int_{\nu_1}^{\infty} \frac{B_\nu(T_*)}{h\nu} d\nu}. \tag{7.6}$$

By summing up the fluxes in all forbidden lines and comparing with the flux in Hβ, we can find T_* by iteration, using Eq. (7.6). The energy balance method was extensively applied to PN by Kaler (1976) and Preite-Martinez and Pottasch (1983). Using the results of the PNe spectroscopic survey of Acker *et al.* (1989), Preite-Martinez *et al.* (1989) and Preite-Martinez *et al.* (1991) were able to determine the central-star temperature for ~500 nebulae. However, it is difficult to include all the fluxes of collisionally excited lines, in particular those in the UV or in the infrared, and collisionally excited HI lines. The stellar temperatures derived by this method can be considered as lower limits.

7.1.3   The Ambartsumyan temperature

In recognition of Ambartsumyan's (1932) observation that the HeII $\lambda 4,686$/ Hβ nebular flux ratio is a reflection of T_*, Kaler and Jacoby (1989) developed this method to estimate the visual magnitudes and effective temperatures of stars embedded in optically thick nebulae. This method uses the nebular emission lines of HeII λ 4,686 Å and H$\beta\lambda$ 4,861 Å to derive the photon fluxes shortward of 228 and 912 Å, and it uses the ratio of these fluxes to obtain a color. Assuming a blackbody spectrum, T_* is iterated until it gives a value of m_V such that $T_{zH}(m_V, F_{H\beta}) = T_{zHeII}(m_V, F_{HeII})$.

7.1.4 Nebular subtraction method

The Zanstra method still requires the measurement of a stellar magnitude (usually m_V), which is difficult for PN with hot central stars because these stars are extremely faint in the visible. Even observing the PN in a passband free of emission lines is not enough because the nebular continuum alone is often brighter than the star. Since the recombination lines and b-f emission depend on nebular density in the same way, a narrow-band Hβ image can be used to trace the nebular continuum emission. Since the star does not contribute to the Hβ image, the nebular contribution can be removed by scaling and subtracting the Hβ image from one taken in a nearby continuum band. Using this technique, one can reveal even very hot central star embedded in bright nebulae (Heap and Hintzen, 1990). With the stellar magnitude measured, the central-star temperature can be derived with the Zanstra method.

7.1.5 Continuum color temperature

With the extension of observations to the ultraviolet, it is possible to measure the stellar continuum over a wide wavelength range. The observed continuum can be compared to a stellar atmosphere model (or a blackbody) to estimate the effective temperature. The strength of the observed λ 2,200 Å extinction bump can be used to estimate the amount of interstellar extinction. The reddening corrected spectrum can then be used to derive a temperature. Figure 7.1 shows the *IUE* low resolution spectrum of the central star of NGC 6543 before and after extinction correction. The reddening-corrected spectrum is fitted by a pure He model atmosphere with $T_* = 80,000$ K and $\log g = 6$ (Bianchi *et al.*, 1986). Although this method makes a direct measurement of the central star, it still suffers from contamination by nebular continuum emission and extinction by circumstellar dust.

7.1.6 Absorption line profiles

The traditional method of using the profiles and fluxes of photospheric absorption lines to deduce stellar temperature is difficult for CSPN because bright nebular emission often fills in or distorts the profiles of stellar absorption lines. Such difficulties can only be overcome by very-high-resolution spectroscopy. Estimates of the temperatures of CSPN have been made by Wilson (1948) and Wilson and Aller (1954), using the equivalent widths of H and He lines and comparing the results with normal O stars.

For central stars that are bright enough to allow the acquisition of high-resolution absorption line profiles, Méndez *et al.* (1988) were able to derive both the effective temperature and the gravity ($\log g$) by analyzing the line profiles in terms of model atmospheres (see Section 7.4.) These results give much more accurate determinations of the central-star temperatures when compared to earlier efforts.

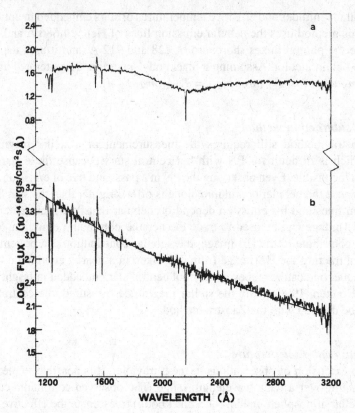

Fig. 7.1. a. The observed spectrum of the central star of NGC 6543 showing the λ 2,200 Å absorption. b. The reddening corrected spectrum obtained by using an extinction law and an E_{B-V} value of 0.08 (from Bianchi *et al.*, 1986).

7.1.7   *Excitation class*

Analogous to the relationship between the stellar spectral types and effective temperatures of main sequence stars, the excitation class is defined to use the nebular lines as a measure of the central star temperature. A scheme based on the excitation of the He lines was first developed by Aller (1956). He classified PN into ten excitation classes (ECs). PN that do not show [OIII] lines are assigned class 1. If the HeII λ 4,686 Å is not present, then the PN is placed in classes 1–5 depending on the [OII] or [OIII] to Hβ ratio. Nebulae that show HeII but not [Ne V] are referred to as class 6. Those that show [Ne V] are classified as classes 7 to 10 depending on the [Ne V] strength.

A continuous EC based on the [OIII] λ 5,007 Å and HeII λ 4,686 Å lines was developed by Dopita and Meatheringham (1990):

$$EC = 0.45(F_{5007}/F_{H\beta}) \qquad 0.0 < EC < 5.0$$

$$EC = 5.54(F_{4686}/F_{H\beta} + 0.78) \qquad 5.0 \leq EC < 10.0 \tag{7.7}$$

Applying a nebular model to a group of Magellanic Cloud PN, Dopita and Meatheringham (1991a, 1991b) determined T_* for these PN and found their values to correlate with the EC defined above. We should note, however, that Kaler and Jacoby (1991)

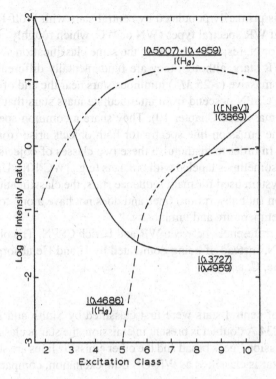

Fig. 7.2. Intensity ratios used to define the excitation class of PN (from Aller and Liller, 1968).

obtained lower temperatures using the energy balance method and found no T_* higher than 70,000 K. For PN with the highest excitation, HeII λ 4,686 Å intensities tend to flatten out and [NeV] may still be needed to determine the excitation class (see Fig. 7.2).

Dopita and Meatheringham (1991b) found a correlation between L_* and EC and express the empirical relationship in a cubic formula:

$$\log(L/L_\odot) = \log(L_{H\beta}/L_\odot) + (2.32 \pm 0.02)$$
$$- (0.179 \pm 0.016)EC + (0.035 \pm 0.004)EC^2$$
$$- (0.00166 \pm 0.00035)EC^3. \tag{7.8}$$

These two empirical formulae [(Eqs. (7.7) and (7.8)] therefore provide a convenient way to transform observed quantities (e.g., Hβ flux) to L_* and T_* for plotting on the H-R diagram (see Chapter 17).

7.2   Spectral classification of the central stars

Spectra of CSPN can be divided into two main groups: those showing H-rich atmospheres and those showing H-poor atmospheres. A more precise scheme was proposed by Méndez (1991), who classifies CSPN based on their surface gravity ("O" for low g and "hgO" for high g) and atmospheric chemical composition (H, He, or C). The ratio of H-rich to H-poor CSPN is approximately 4 to 1, similar to the ratio observed in WDs (see Section 15.3).

The group of H-poor CSPN is primarily populated by central stars with Wolf-Rayet (WR) spectra. Stars are assigned WR spectral types (WN or WC, which roughly correspond to objects defficient in C or N, respectively) using the same classification system as that used for Population I WR stars, although they are fundamentally different objects. Population I WR stars are massive (>25 $M_\odot$) luminous stars near the end of their nuclear burning phase, whereas CSPN descend from intermediate mass stars that have a very low (~0.6 $M_\odot$) current mass (see Chapter 10). They share a common spectral classification system because the emission-line spectra for both objects arise from an expanding H-poor stellar wind. In order to distinguish these two classes of objects, the WR spectral types of CSPN are sometimes labeled with brackets (e.g., [WC 10]). Unlike the MK spectral classification system used for main sequence stars, the classification of WR spectra is based on emission (not absorption) lines and does not have a one-to-one correspondence with effective temperature and luminosity.

Figure 7.3 shows the spectral differences between WR and H-rich CSPN. The photospheric continuum of WR CSPN, instead of being dominated by H and He absorption lines, shows emission lines of He, C, and O.

7.2.1   WR central stars

Emission-line spectra of central stars were first classified by Smith and Aller (1969). If the OVI λ 3,811 Å/3,834 Å doublet is present in emission, the star is classified as OVI. If the HeII λ 4,868 Å emission line is broad and the other emission lines are sharp, like those of Of stars, then the star is classified as WR-Of. Low-excitation, compact PN

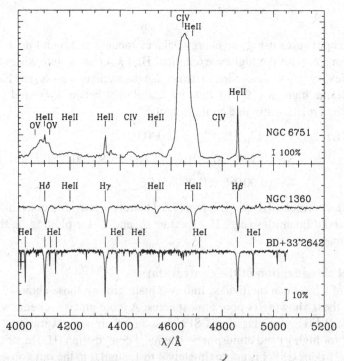

Fig. 7.3. Spectra of WR (top), hot ($T_{\mathrm{eff}} \sim$ 110,000 K) H-rich (middle), and cool ($T_{\mathrm{eff}} \sim$ 20,000 K) H-rich (bottom) CSPN (figure from Napiwotzki, 1998).

with WR-type nuclei were discovered by Webster and Glass (1974). This has led to an extension of the WR sequence to types later than those of massive WR stars. Using the results from the spectroscopic survey conducted for the Strasbourg-ESO catalog of Galactic Planetary Nebulae, Tylenda *et al.* (1993) classified 39 as WR and 38 as "weak emission lines." The latter class refers to CSPN with the CIV λ 5,805 Å-line weaker and narrower than early [WC] stars. The WR central stars are all of the [WC] type, with concentrations in the [WC 3-4] and [WC 9-11] subclasses. The temperatures range from as low as 20,000 K for [WC 11] stars to as high as 150,000 K for the [WC 2-3] stars. There are no well established cases of [WN] central stars. Górny and Stasińska (1995) estimate that approximately 8% of all PN have WR central stars.

The association of H-deficient central stars with nebulae of young dynamical age is particularly interesting (Kwok *et al.*, 1993; Zijlstra and Siebenmorgen, 1994). For example, BD + 30 3639, a [WC 9] star, has a dynamical age of only 900 yr. If the H deficiency in this object is due to a late He flash while the star is on the cooling track, then the star must be relatively massive ($\sim$0.8 $M_\odot$) in order to evolve fast enough to get there (see Section 11.1). Whether all [WR] central stars are of higher core mass is still uncertain and the subject of continued debate.

7.2.2 Infrared properties of WR central stars

Interestingly, PN with late-type WR central stars have much greater infrared excesses [IRE; see Eq. (6.18)] than PN with non-WR central stars. This suggests that all the photons longward of the Lyman limit are absorbed by circumstellar dust and are converted to far infrared photons. The most extreme case is IRAS 21282+5050, a PN with a [WC 11] central star, which has an IRE value of $\sim$300 (Fig. 7.4). IRAS 21282+5050 is one of several PN with [WC 11] nuclei that show strong UIR emission features in the near infrared (Table 7.1). Their infrared spectra are very similar to that of NGC 7027, a young PN with a massive central star. They are also reminiscent of NGC 7027 in that they often possess molecular envelopes. It is likely that these [WC 11] stars are

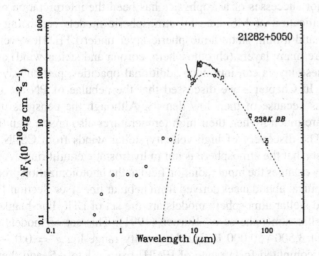

Fig. 7.4. The SED of IRAS 21282+5050 showing the very strong IR excess in this object (from Kwok *et al.*, 1993).

Table 7.1. *Cool WC stars with large infrared excesses*

PN	Spectral Type	UIR features (μm)
M4-18	[WC 11]	8.6, 11.3
Vo 1	[WC 11]	7.7, 11.3
He2-113	[WC 11]	3.3, 7.7, 8.6, 11.3
CPD-56°8032	[WC 11]	3.3, 7.7, 11.3
IRAS 21282+5050	[WC 11]	3.3, 6.2, 7.7, 8.6, 11.3

Table taken from Kwok *et al.*, (1993).

carbon-rich, high-mass stars that have evolved rapidly across the H-R diagram. On the other hand, the WR and Of-WR central star samples have the same galactic distribution as all PN, giving no evidence that they are formed from a more massive subgroup of progenitors (Pottasch, 1996).

In spite of the fact the [WC] stars have carbon-rich central stars and show carbon-rich PAH features, crystalline silicate features have been detected in a significant fraction of [WC] stars, suggesting that oxygen-rich envelopes are still present in these objects (Waters *et al.*, 1998).

7.3   PG 1159 stars

Schönberner and Napiwotzki (1990) showed that several CSPN are spectroscopically indistinguishable from the PG 1159-035 WDs (see Section 15.5). Napiwotzki and Schönberner (1991) identified several CSPN that show Balmer absorption in addition to their characteristic PG 1159 features. These stars are assigned the classification O(H,C) by Napiwotzki (1993).

7.4   Model atmospheres

One of the major successes of astrophysics has been the interpretation of the empirical spectral classification of MS stars by atmospheric models consisting of a blackbody photosphere and a thin, static atmospheric layer under LTE. However, we now realize that there are many layers (chromosphere, corona and stellar wind) above the photosphere, and these layers can introduce additional opacities, particularly outside the visible region. In Chapter 3 we discussed that the nebulae of PN are often far from LTE conditions because of their low density. Although the densities in the atmospheres of CSPN are much higher, their high temperatures also create significant departures from LTE. The discovery of high-velocity stellar winds from CSPN (see Section 7.5) also suggests that the atmosphere is not in hydrostatic equilibrium. A non-LTE atmosphere not only changes the input radiation field to the photoionization process, it also can affect the chemical abundances derived from nebular lines (see Section 19.1).

The most widely used stellar atmosphere models are the set of LTE, line-blanketed, static, plane-parallel stellar atmospheres by Kurucz (1991). This set of models span the temperature range of 3,500−50,000 K and the gravity range log $g = 0.0 − 5.0$. The Kurucz models are computed for a range of [Fe/H] from +1 to −5, and they include up-to-date line opacities and continuous opacities. Non-LTE treatments of CSPN

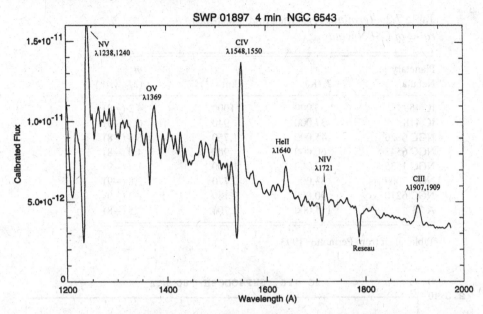

Fig. 7.5. The *IUE* spectrum of NGC 6543. Four lines can be seen showing the P Cygni profile. The vertical scale is in units of erg cm^{-2} s^{-1} Å^{-1} (from Feibelman, 1994).

include Werner and Heber (1991), Clegg (1987), and Husfield *et al.* (1984). These models cover the T_e − log g grids of 80,000−200,00 K and 5.0−8.0 (Werner and Heber, 1991), 40,000−180,000 K and 4.0−8.0 (Clegg and Middlemass, 1987), and 50,000−115,000 K and 3.8−5.75 (Husfield *et al.*, 1984). More recently, expanding, non-LTE model atmospheres including millions of metal lines have been constructed (Kudritzki *et al.*, 1997). It is shown that the emergent flux distribution deviates significantly from a blackbody, and metal-line blanketing has potential effects on nebular diagnostics. If this is the case, then the input radiation field used in nebular photoionization models should be replaced by more realistic stellar atmosphere models.

7.5 Winds from the central star

Broad lines in the spectra of CSPN implying velocities of ∼1000 km s^{-1} were noted as early as 1945 by Minkowski. The spectral classification of central stars by Smith and Aller (1969) show that many central stars are WR stars with P-Cygni profiles. However, the realization of the common occurrence of high-velocity winds from CSPN only came after the observations of UV resonance lines from the *IUE* satellite (Heap *et al.*, 1978; see Figs. 7.5 and 7.6). Further observations with the *IUE* show that ∼60% of central stars with measurable continuum show evidence of winds (Cerruti-Sola and Perinotto, 1985; Patriachi and Perinotto, 1991). The observed terminal velocities are 1,400−5,000 km s^{-1}, or approximately several times the escape velocities of the central stars (Table 7.2). The wind velocities can be determined within ±200 km s^{-1}, with still-better accuracies possible with high-resolution *IUE* spectra.

The mass-loss rates of the central-star winds can be derived by comparing the observed P-Cygni profiles with theoretical profiles constructed assuming a velocity law and

Table 7.2. *Terminal velocities and mass-loss rates of CSPN winds*

Planetary Nebula	$T_*(\mathrm{K})$	v (km s^{-1})	$\dot{m}$ ($M_\odot$ yr^{-1})
IC 4593	35,000	1000	4.2 (−8)
IC 418	37,000	940	6.3 (−9)
NGC 6826	45,000	1,750	6.4 (−8)
NGC 6543	60,000	1,900	4.0 (−8)
NGC 1535	77,000	1,900	1.4 (−9)
NGC 7009	88,000	2,770	2.8 (−9)
NGC 6210	90,000	2,180	2.2 (−9)
A 78	115,000	3,700	2.5 (−8)

Table taken from Perinotto (1993).

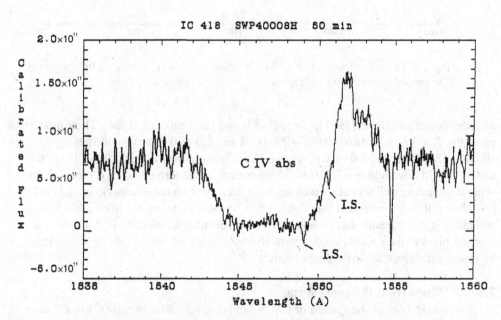

Fig. 7.6. High resolution *IUE* spectrum of NGC 40 showing the P-Cygni profile of the C IV line. A wind terminal velocity of 1730 ± 50 km s^{-1} is derived from this spectrum (from Fibelman, 1999).

ionization structure. For an accelerating stellar wind, the optical depth at each frequency is appreciable only over a narrow range in radius where that frequency is matched by the Doppler-shifted rest frequency of the line. This is known as the Sobolev approximation and an expression for the radial optical depth is given by Castor (1970):

$$\tau(r) = \frac{\pi e^2}{m_e c} f_{12} \lambda_0 \left(\frac{dr}{dv}\right) \left(n_1 - \frac{g_1}{g_2} n_2\right), \tag{7.9}$$

where λ_0 is the rest wavelength of the transition, n_1 and n_2 are the ion number density in the lower and upper levels at distance r, and dr/dv is the inverse velocity gradient. If the velocity increases monotonically with radius, dr/dv is uniquely defined at each radius. Therefore $\tau(r)$ is the optical depth presented by the envelope to a photon traveling radially that would be resonantly absorbed at radius r.

Assuming $n_1 \gg g_1 n_2/g_2$ and letting $x = r/R_*$ and $w = V(r)/v_\infty$, we find that the integrated optical depth over the profile is given by

$$\tau_{\text{tot}} = \int_0^1 \tau \, dw. \tag{7.10}$$

Substituting Eq. (7.9) into Eq. (7.10), we have

$$\tau_{\text{tot}} = \frac{\pi e^2}{m_e c} \frac{f_{21} \lambda_0 R_*}{v_\infty} \int_1^\infty n_1 \frac{dx}{dw} dw$$

$$= \frac{\pi e^2}{m_e c} \frac{f_{21} \lambda_0}{v_\infty} \int_{R_*}^\infty n_1 dr. \tag{7.11}$$

The product $\tau_{\text{tot}} v_\infty$ is proportional to the column density and can be used as a curve-of-growth parameter.

Assuming a constant mass-loss rate and constant excitation and ionization

$$n_1 = \left(\frac{n_1}{n_A}\right)\left(\frac{n_A}{n_H}\right)\left[\frac{\dot{M}}{4\pi \mu m_H v(r) r^2}\right], \tag{7.12}$$

where n_A is the total number density of element A. Substituting Eq. (7.12) into Eq. (7.11) gives

$$\tau_{\text{tot}} = \left(\frac{\pi e^2}{m_e c}\right)\left(\frac{f_{12} \lambda_0}{R_* v_\infty^2}\right)\left[\frac{\frac{n_1}{n_A} \frac{n_A}{n_H} \dot{M}}{4\pi \mu m_H}\right] \int_1^\infty \frac{dx}{x^2 w(x)}. \tag{7.13}$$

A simple analytical expression for the velocity field is used by Castor and Lamers (1979)

$$w(x) = 0.01 + 0.99\left(1 - \frac{1}{x}\right)^\beta. \tag{7.14}$$

For $\beta > 0$, Eq. (7.14) ensures that $w = 1$ at large distances from the star and w is small at the photosphere. For example, Lamers and Morton (1976) used the $\beta = 0.5$ law to fit the profiles of ζ Pup.

For P-Cygni profiles with almost equal absorption and emission components, the first moment of the profile is close to zero. The second moment, defined as

$$W_1 = \left(\frac{c}{\lambda_0 V_\infty}\right)^2 \int \left(\frac{F_\lambda - F_{\text{cont}}}{F_{\text{cont}}}\right)(\lambda - \lambda_0) \, d\lambda, \tag{7.15}$$

is the most convenient to use to compare with theoretical profiles. For optically thin lines, W_1 is directly related to the τ_{tot}. By assuming that

$$W_1 = 0.30 \times \tau_{\text{tot}} \tag{7.16}$$

and approximating the integral in Eq. (7.13) with a value of 2, Castor *et al.* (1981) derive the following expression:

$$\dot{M} = 2 \times 10^{-18} \frac{(R_*/R_\odot)(v_\infty/\mathrm{km/s})^2 W_1}{\frac{n_1}{n_A}\frac{n_A}{n_H} f_{12}(\lambda_0 v)} M_\odot \mathrm{yr}^{-1}, \tag{7.17}$$

which is widely used (e.g., by Cerruti-Sola and Perinotto, 1985) to derive mass-loss rates of CSPN.

The mass-loss rates derived are uncertain because of problems associated with the excitation and ionization states of the atoms observed (Castor *et al.*, 1981). This is further complicated by the fact that central stars of PN have nonsolar composition in the atmosphere because of previous stages of nuclear shell burning (see Section 19.1). Mass loss rates obtained from the fitting of the line profiles range from 3×10^{-10} to 10^{-7} $M_\odot$ yr^{-1} (Table 7.2).

The wide presence of high-speed winds from the CSPN implies that the interaction between this wind and the nebula is inevitable. The mechanical energy input provided by the central-star wind is also higher than the kinetic energy associated with the observed expansion of PN. The effects of these winds on the formation and the dynamical evolution of PN will be discussed in Section 12.2.

7.6   Extreme UV and X-ray emission from CSPN

The high temperatures of CSPN suggest that they are likely sources of extreme UV (EUV) or soft X-ray sources. X-ray and EUV observations of PN using *Einstein* (Tarafdar and Apparao, 1988), *EXOSAT* (Apparao and Tarafdar, 1989), *ROSAT* (Kreysing *et al.*, 1993), and the *Extreme Ultraviolet Explorer* (*EUVE*, Fruscione *et al.*, 1995) satellites have identified a number of point sources with PN. These high-energy radiations can be generated either by photospheric continuum emission from the central star or by high temperature gas in the stellar coronal regions.

Spectral fitting to the X-ray data generally yields blackbody color temperatures that are higher than the effective temperatures obtained by other methods. This is likely caused by the nonthermal nature of the photospheric spectrum.

7.7   PN with binary central stars

Close-binary PN central stars can be discovered by (i) photometric variability; (ii) raidal velocity variations (e.g. in NGC 2346); and (iii) composite spectra (e.g., Abell 35). Wide (visual) binaries can be found by high-spatial-resolution observations. Approximately 15 PN have been found to have close-binary nuclei, with orbital periods ranging from 2.7 hr to 16 days (Bond, 1995). A list of PN with known close-binary central stars is given in Table 7.3.

Because of the possibility of interaction and mass transfer, binary systems can have a very complicated evolutionary path both before and after the PN phase. For example, for close binaries the primary component can fill its Roche lobe during the AGB and the following mass transfer can result in a common envelope phase (Iben and Livio, 1993).

Table 7.3. *Planetary nebulae with close-binary central stars*

Planetary Nebula	Central Star	Period (days)	Binary Type
Abell 41	MT Ser	0.113	Reflection
DS 1	KV Vel	0.357	Reflection
Abell 63	UU Sge	0.465	Eclipsing
Abell 46	V477 Lyr	0.472	Eclipsing
HFG 1	V664 Cas	0.582	Reflection
K 1-2	VW Pyx	0.676	Reflection
Abell 65	⋯	1.00	Reflection
HaTr 4	⋯	1.71	Reflection
(Tweedy 1)	BE UMa	2.29	Eclipsing
Sp 1	⋯	2.91	Reflection
SuWt 2	⋯	4.8	Eclipsing
NGC 2346	V651 Mon	15.99	Spectroscopic
Abell 35	−22° 3467	⋯	IUE composite
LoTr 1	⋯	⋯	IUE composite
LoTr 5	HD 112313	⋯	IUE composite

Table taken from Bond (1995).

7.7.1 Relationship between symbiotic stars and PN

Symbiotic stars are binary systems consisting of a hot WD and a late-type star surrounded by an ionized nebula. The nebular spectrum consists of highly excited emission lines similar to that of a PN spectrum. The major observational distinction between a PN and a symbiotic star is the presence of a photospheric continuum that is due to the cool component of the symbiotic system. The presence of a late-type star is confirmed if the continuum shows molecular absorption features such as TiO and VO.

The similarities between PN and symbiotic novae (e.g., V1016 Cyg and HM Sge) are particularly striking: symbiotic novae often show evidence of a nebular shell created by the collision of a fast wind and a slow wind, a process which is responsible for the formation of PN (see Section 12.2). The only difference is that the two winds in a symbiotic system are supplied by two different stars: the fast wind from the hot component and the slow wind is from the AGB component (Kwok, 1988). Since the hot component is likely to have gone through a PN phase before it is reignited by mass transfer, and the cool component, a mass-losing AGB star, will also evolve to a PN on its own in a million years, the symbiotic novae could represent an interlude between two PN phases! Furthermore, in the second PN phase, the nebula will be seen to have two central stars: one more luminous (the core of the present AGB star) and one WD on the cooling track.

7.8 Summary

CSPN have high temperatures and emit most of their fluxes in the UV. The photospheric continua are often masked by nebular continuum emission. These two factors together make the effective temperatures of CSPN difficult to determine. The most commonly used techniques rely on the nebular properties to derive the central temperatures and this can often lead to errors (see Section 13.7).

Unlike MS stars, CSPN cannot be classified by a one-dimensional scheme because of their inhomogeneous chemical compositions. The photospheric chemical compositions of MS stars reflect metallicity of the environment from which they were born, but the photospheric compositions are affected by effects of evolution.

The spectra of CSPN are made more interesting (and complicated) by the presence of stellar winds. Since CSPN have very thin H and He envelopes, these stellar wind accelerate the depletion the envelope and hasten the evolution of the CSPN (see Section 11.2). Depending on the exact mass-loss formula (which is poorly known), CSPN can evolve simply as H burners, or undergo wild changes in the evolutionary tracks caused by the ignition of He (see Section 11.4). Furthermore, the mechanical energy carried by these winds is responsible for the dynamical expansion of the nebula (see Chapter 12).

8

Morphologies of planetary nebulae

Although PN are well known for their ring-shape appearance, they in fact have a diverse range of morphologies. Using photographs that he took at the Lick Observatory, Curtis (1918) was the first to arrange PN into different classes based on their appearances. The origin of such diverse shapes has remained a mystery for a long time. For example, the well known Ring Nebula (NGC 6720) has an elliptical ring appearance. The most obvious interpretation is that this represents a three-dimensional hollow shell projected onto the sky. However, the actual observed surface brightness of the "hole" in comparison to the shell is too low ($\sim 1:20$) to be consistent with this model (Minkowski and Osterbrock, 1960). The observed intensity distribution is in fact more compatible with an open-ended toroid viewed end on (Khromov and Kohoutek, 1968). Although this model gives a good approximation to the observed image, the origin of such a toroid is not explained. The physical origin of the different morphologies of PN and how they evolve to such forms therefore represents one of the greatest challenges in PN research.

8.1 Morphological classifications

Curtis (1918) classified his sample of 78 PN into helical, annular, disk (uniform and centrally bright), amorphous, and stellar. Subsequent classification schemes often use similar descriptive forms: stellar, disk, irregular, ring, anomalous (Perek & Kohoutek, 1967); elliptical, rings, bipolar, interlocking, peculiar, and doubtful (Greig, 1971; Westerlund and Henize, 1967); and round, elliptical, and butterfly (Balick, 1987). Two hundred and sixty three photographic images in the Perek and Kohoutek catalog were classified by Khromov and Kohoutek (1968) into three main types based on their main structures and several subtypes based on the peripheral structures. They emphasized the role of projection effects on the observed structures of PN and concluded that many PN have toroidal structures in spite of their apparent forms. Greig (1971) studied 160 PN from various sources and correlated their morphology to other characteristics of the objects, such as type of central star and nebular excitation. He concluded that there are principally only two major groups of PN – class B and class C. Class B PN have large ansae, tubular, or filamentary structures. Many class B members that are bipolar are brightest at the ends of, or along, the minor axis with a sharp cut-off in brightness beyond the minor axis. Furthermore, this type often has strong forbidden lines of [NII], [OII], and [OI] relative to the Balmer lines; many of the nuclei are of type [WC] and they seem to be located at a significantly lower mean galactic latitude than the other type. Class C PN have a centric increase in brightness and they are more smooth in appearance than class B. They have weaker forbidden-line intensities and are located at significantly higher mean galactic latitudes. The above observations suggest that the morphology of PN may be related to other fundamental properties of the central star.

Westerlund and Henize (1967) carried out an examination of 151 Hα images of PN and concluded that young objects (small and bright) showed greater structural regularity. They suggest that this could represent either a resolution effect or an evolution effect.

Two hundred and fifty five Hα and [OIII] ($\Delta\lambda = 50$–75 Å) images of PN were obtained by Schwarz *et al.* (1992). Because of the higher dynamic range of the CCD images, faint halos are often seen. The angular sizes of the observed nebulae ranged from $\sim$1 arcsec to 4 arcmin. These images were classified by Stanghellini *et al.* (1993) into the following classes: (i) stellar, (ii) elliptical, (iii) bipolar, (iv) point symmetric, and (v) irregular. They define bipolar as nebulae that show a "waist." A catalog of CCD images of PN can be found in Manchado *et al.* (1996).

Using the radio images obtained in the VLA surveys, Aaquist and Kwok (1996) classified these images in terms of the ellipsoidal shell model and divided them into circular, elliptical, open elliptical, and butterfly types. Their butterfly class is equivalent to the bipolar class of Stanghellini *et al.* (1993). Two examples of this type of PN is shown in Figs. 8.1 and 8.2. Those showing asymmetry are classified into S, peak excess,

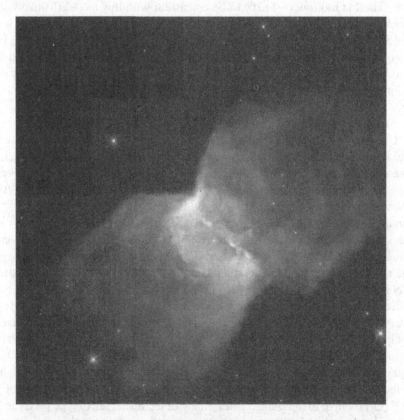

Fig. 8.1. Example of a "butterfly"-class PN. The central star of NGC 2346 is a spectroscopic binary with a period of 15.99 days (see Section 7.7). This HST WFPC2 image was taken during the recommisioning of the HST after the first service mission.

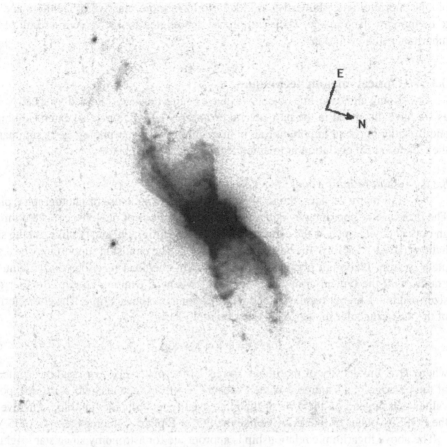

Fig. 8.2. The *HST* WFPC2 [Nɪɪ] image of NGC 6881.

blister, unbalanced poles, symmetric, halo, and ansae types. Aaquist and Kwok (1996) concluded that PN took their shape early in their evolution. Statistically, their sample of young PN was not different in shape from samples of more evolved PN.

8.2 Relationship between nebular morphology and central star evolution

Stanghellini *et al.* (1993) studied the location of the central stars of nebulae with different morphological types on the H-R diagram. They proposed that, for example, bipolar and elliptical PN very likely contain central stars with a different mass distribution. From a sample of 43 bipolar nebulae, Corradi and Schwarz (1995) suggested that bipolar nebulae have more massive progenitors. However, descriptive classifications of apparent morphologies can be misleading because the projection effect cannot readily be disentangled from other fundamental, intrinsic properties. In fact, some of the best-studied PN such as NGC 6720 (the Ring Nebula) and NGC 7027 will almost certainly have bipolar appearances if they are viewed edge on. Zhang and Kwok (1998a), using the ellipsoidal shell model to derive the orientation angle and the angular density gradients of 110 PN, were able to remove the projection effects. They found that PN of extreme bipolar morphologies (butterfly) are likely to have more massive progenitors and also to be more evolved in age. Their extreme morphologies are therefore the result of both inheritance and evolution.

8.3   Optical imaging techniques

Being emission-line objects, PN have been extensively studied by spectroscopy, as we have discussed in the past several chapters. Many PN are also extended objects, and imaging plays an important role in determining their morphologies. A summary of the technological evolution in imaging techniques is given below.

8.3.1   *Photographic plates*

For many decades, the imaging of PN relied on the use of photographic plates. The principle of photography is based on the interactions of photons, collected through an optical device such as the camera, telescope, spectrograph, and the like, on the silver halides (BrAg) that form the basis of the emulsion. The emulsion support may be a glass plate or film. Incoming photons provoke random chemical reactions on the emulsion grains, with the typical grain size varying between 25 μm and 75 μm. Smaller grain sizes result in a higher resolution and require a longer exposure time. The characteristics of the stellar/nebular images obey the reciprocity law:

$$\log D = p \log t + \log E, \tag{8.1}$$

where D is the density of the image stored on the plate, E is the brightness/intensity of the photographed source, t is the exposure length in seconds, and p is the exponent values that depend on the type of the plate used (panchromatic, i.e. blue sensitive like the POSS-O plates or red sensitive like the POSS-E plates). In practice, $p \approx 0.85$.

The above logarithmic relationship is appropriate for astronomy since star brightness is scaled in magnitudes to fit the human eye visibility curve, which peaks at approximately λ 5,500 Å. In practice, optical astronomers characterize a given photographic plate by plotting logarithms of D(ensity) against logarithms of various values of E by illuminating the photographic plates with various intensity levels (either absolute or relative). In order to allow photometric use of the recorded data (spectra and images), photographic plates must be calibrated. This is done by illuminating the photographic plate with an artificial light source. For spectroscopic purposes, a emitting line source (Hg, Ar, Cd, etc.) is needed to calibrate the spectrograph in the desired wavelength intervals. This operation defines the dispersion function of the spectrograph in angstroms per millimeter. For photometric measurements, a continuum source, usually an incandescent (tungsten) lamp, is used to illuminate the whole device through a series of diaphragms whose diameters vary in known ratios. This determines the response of the photographic plate to an input signal within a given dynamical range. In practice, the plot $\log D$ versus $\log E$ has an S shape, the lower the brightness the lower the density of the plate image. The saturation corresponds to the asymptotic region of the S (high density) curve. Data are usually recorded in the linear region of such a sensitivity curve, which allows a rather limited dynamic range: $\Delta \log D \sim 3-4$ magnitudes.

Although photographic plates are mainly characterized by a "clumpy" (random) structure of the BrAg emulsion, such devices have provided a tremendous quantity of photometric data in the past. In the blue section of the spectrum, astronomers used either Kodak II aO (moderately high-sensitivity because of moderate grain sizes) or 103 aD (less rapid but allowing a higher resolution because of smaller grain sizes). In the red spectral region, pre-nitrogen-heated 1N plates/films were often used. Because of its restricted dynamical range, it is difficult to observe celestial objects with large differences in brightness, for example, galaxies with bright nucleus and very faint spiral arms and PN (such as NGC 6543, NGC 40) displaying a high contrast between the central areas and the outermost emissions.

The advent of image tubes helped to extend the observations toward the spectral ranges to which photographic plates are not directly sensitive, because of the fluorescence mechanism: UV photons hit a phosphorous medium which in turn will emit visible photons to impress the photographic plates.

8.3.2   The image photon counting systems

Image photon counting systems (IPCS) were developed in the early 1970s, and their principle of operation is to detect individual photons that are multiplied by means of a photomultiplier-amplifier-intensifier. This differs from photographic plates, which have to receive groups of photons. Another important improvement made possible by this new device is that one no longer has to wait before developing the plate to know what has actually been acquired; the observation can be made in real time and observers can visualize the acquisition simultaneously while the telescope is tracking. Since its response is linear (Poisson law signals), faint details can be imaged with this device.

8.3.3   The charge-coupled devices

The IPCS available during the period 1970–1985 provided a relatively limited field of view since the largest array had 256×256 pixels of $\sim 30~\mu$m. For example, at the Cassegrain $f/8$ of the Canada-France-Hawaii 3.6 m telescope, the INSU-IPCS offered a field of view of ≈ 1 arcmin $\times$ 1 arcmin. This changed with the development of CCDs.

The first alternative detector based on a charge transfer mechanism, developed at the Bell Labs in the early 1970s had a rather limited format. It had a size of 100×100 pixels and a sensitivity even lower than some good photographic plates in use at the time. The performance of a CCD detector is dependent on the fluctuations it adds to those of the signal (called detector quantum efficiency) and the read-out noise, which is measured in number of electrons per reading. Nowadays, after the historical 320×512 RCA chip, astronomers can count on 2k $\times$ 2k, thin, back illuminated, low read-out noise, high detector quantum efficiencies, and small pixel size (~ 10–$15~\mu$m).

Like the IPCS, CCDs are characterized by a linear response within the limit of saturation. The typical quantum efficiency of a CCD is relatively high between 4,000–8,000 Å, but drops off quickly at wavelengths shorter than 4,000 Å. The usual operational temperature is $-110°$ C, although high performance CCDs have been made with cooling temperature between $-40°$ C and $-80°$ C. Because of the linearity of the signal, absolute calibrations are easy to obtain, resulting in good-quality photometry.

The development of these new technology devices has revolutionized the imaging of PN. New faint outer structures were discovered around many PN, even among well-observed objects such as the Ring Nebula (NGC 6720), Dumbbell (NGC 6853), NGC 40, and so on.

8.4   Halos around planetary nebulae

Faint halos outside the main shells of PN have been known since 1937 (Duncan, 1937; Minkowski and Osterbrock, 1960; Milikan, 1974; Kaler, 1974) and are traditionally interpreted as the result of multiple PN ejections (Capriotti, 1978). The use of CCD technology has greatly improved the dynamic range of optical observations, and many previously unseen faint outer structures since have been revealed. In the CCD survey of PN by Jewitt *et al.* (1986), two-thirds of the PN observed were found to possess extensive outer halos. The halo material usually appears to be contiguous with the main shell and has a density of ~10% of that of the main shell. The extent of the halo has been seen to as far as five times the radius of the shell (Hua, 1997). These developments suggest that halos around PN may be a very common, if not universal, phenomenon.

Morphological studies of PN halos by Chu *et al.* (1987) suggest that they can be classified into two types: Type I halos have faint and filamentary material detached from the primary shell, and Type II halos show bright and amorphous structures attached to the main shell. High resolution echelle observations of the halos indicate that Type I halos are often nearly static, whereas Type II halos are expanding at velocities similar to those of the primary shell. Most interestingly, some elliptical PN have been found to have a spherical halo (Fig. 8.3), suggesting that the elliptical morphology develops after the AGB phase.

Using the images of Schwarz *et al.* (1992), Stanghellini and Pasquali (1995) classified the halo morphologies as (i) detached halo, multiple-shell when there is a pronounced minimum in the surface brightness profile between the inner nebula and the outer rim of the halo; (ii) attached halo, multiple shell when the halo brightness fades gradually from the bright inner nebula toward the outer edge; or (iii) irregular attached-halo when the halo shows irregularities and high ellipticity. Detached halos of PN were first explained as multiple mass-loss events on the AGB if they have H-rich central stars; or as the result of He-shell flashes on the AGB (with the inner shell representing the remnant of the final flash) if the central star is H depleted. A discussion on the origin of the halos is given in Section 13.6.

8.5   Microstructures

Microstructures are small structures found in clumps near the outer edges of rims of PN. They were first detected as a result of their extinction or because their ionization was lower than the rest of the nebula. These microstructures might have formed as the result of instabilities.

8.5.1   FLIER

Some PN show a pair of small (~0.01 pc), bright knots of low-ionization gas along their major axes. These knots, sometimes known as ansae, are mostly found in elliptical nebulae (Fig. 8.4). Spectroscopic observations have shown that these knots are in low ionization states (primarily seen in lines such as [NII] and [OI]), and have highly

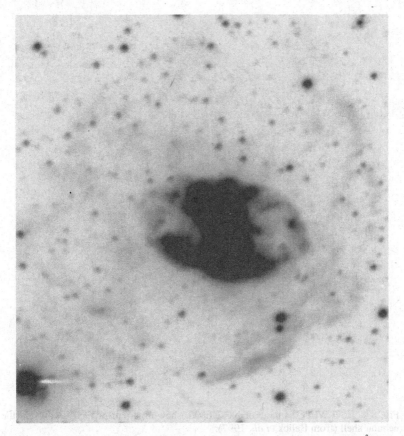

Fig. 8.3. Example of a halo surrounding the PN shell. This is a $\Delta\lambda = 10$ Å [NII] image of He 2-119 (from Hua *et al.*, 1998).

supersonic velocities ($\sim$50 km s^{-1}). They have been named FLIER (fast low-ionization emission regions) by Balick *et al.* (1994). The origin and the acceleration mechanism of FLIER are not well understood.

8.5.2 Cometary knots

Cometary knots are dark objects accompanied by a luminous cusp. They represent a higher concentration of matter (in particular dust) and appear in silhouette against the background of nebular emission. The tails trailing away from the cusps lie closely on radial lines passing between the cusp and the central star. Their appearances therefore resemble that of cometary tails in the solar system. The cusps are probably photoionized surfaces on the face of a neutral core. A radial-outward force is needed to explain the streaming out of material in the tail of the cusps. However, radiation pressure (as in the case of comets) does not seem sufficient.

Cometary knots are best seen in NGC 7293 (the Helix nebula), an old PN that is also probably the closest PN to earth. The HST WFPC2 image of the Helix nebula obtained by O'Dell and Handron (1996) indicates that there are $\sim$3,500 cometary knots in the entire nebula. Assuming an average density of 10^6 cm^{-3}, each cometary knot would have

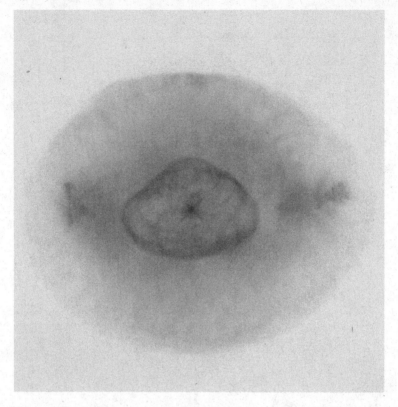

Fig. 8.4. HST WFPC2 image of NGC 6826, showing a pair of FLIERs outside the main nebular shell (from Balick *et al.*, 1997).

a mass of $\sim 3 \times 10^{-5} \ M_\odot$. This suggests a total of $\sim 0.1 \ M_\odot$ of mass in knots in the entire nebula.

Compact H-poor knots with wind-blown tails have also been observed in Abell 30 (Borkowski *et al.*, 1995). The knots lie on spoke like radial lines from the central star. These dense clumps of material are being photo evaporated by the stellar radiation and accelerated by the stellar wind. The dynamical time for the ejection of these material is ~ 1000 yr.

8.5.3   Point-symmetric structures

Point-symmetric pairs of knots in an S-shape structures, or sometimes referred to as bipolar, rotating, episodic jets (BRET), have been seen in a number of PN (López *et al.*, 1995). Colliminated outflows ("jets") can be seen in two of the corners of the [NII] image of NGC 6543 (Fig. 8.5).

8.6   Origin of PN morphology

In order to develop a physical understanding of the morphology of PN, we first have to confront some of the problems with existing morphological classification schemes:

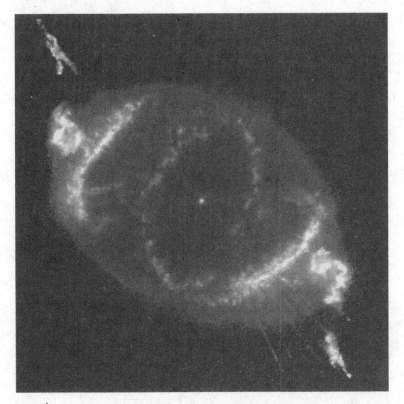

Fig. 8.5. Jets can clearly be seen in this HST WFPC2 [NII] image of NGC 6543.

- Sensitivity dependent: The use of high-dynamic-range detectors also led to the discovery that the morphology of PN may depend on how deep the exposure is. Figure 8.6 shows the images of SaWe3. In a short exposure, only the bright core in the center can be seen. The faint outer structure revealed in a long exposure in fact has the major axis perpendicular to the major axis of the core, which actually represents the "waist" of the equatorial disk of a bipolar nebula. This example shows that morphological classifications can be misleading.
- Species dependent: The morphology of PN observed in lines of different ions is not necessarily the same as the result of ionization structures and stratification effects. The theory of the stratification structures of PN began with the photoionization work of Strömgren (1939). Given the different ionization potentials of different atoms and ions, the spatial distributions of different ions can be quite different in the same nebula (Seaton, 1954a; Köppen, 1979). Possible stratification effects have been observed in PN, although the effects are not as strong as predicted by the spherically-symmetric, homogeneous models (Czyzak and Aller, 1970). When the images of PN are separated by narrow interference filters, differences in morphological structure often appear. For example, when the Helix Nebula (NGC 7293) is imaged in Hα and [NII] λ 6,583 Å, two lines that normally would have been imaged together in a wide-band image,

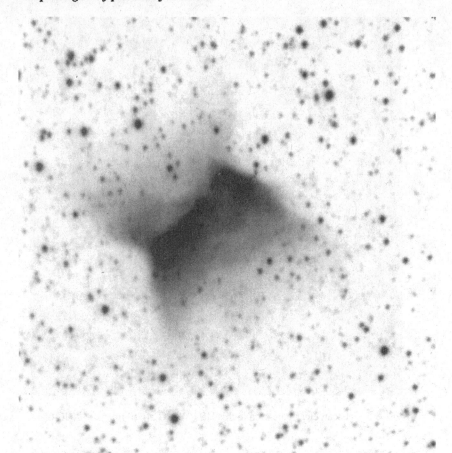

Fig. 8.6. The [NII] image of SaWe3 (PNG013.8-02.8, from Hua *et al.*, 1998).

morphological differences can be seen (Carranza *et al.*, 1968). A program of monochromatic imaging of PN was started by Hua and Louise (1970) using selective interference filters with $\Delta\lambda \leq 10\,\text{Å}$ in order to quantitatively measure the contribution of each ion. Figure 8.7 shows the Hα and [NII] images of the Ring Nebula (NGC 6720). We can see that the ring is filled in in the Hα image but not in the [NII] image. Outer filamentary structures can also be seen in the [NII] image.

• Absolute calibration: The published radio images almost always show an absolute calibration scale. Since f-f emission is only weakly dependent on T_e, the observed intensity distribution can be readily interpreted in terms of the ionized gas distribution. In comparison, the optical atlases often do not have an intensity scale, and therefore it is difficult to translate the images into emission measure (Pottasch, 1995). A quantitative model can only be developed if the image is absolutely calibrated, as it is done, for example, for the NGC 3587 (the Owl Nebula) by Minkowski and Aller (1954) and for IC 351 by Hua and Grundseth (1986).

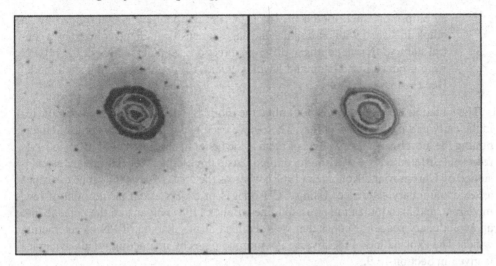

Fig. 8.7. The Hα (left) and [NII] (right) images of NGC 6720 (from C. T. Hua).

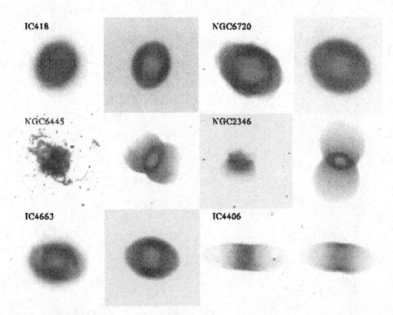

Fig. 8.8. Comparison between the Hα images of six PN and simulated images using the ellipsoidal shell model.

- Morphology classifications give apparent structures, not intrinsic structures. The morphological classification describes the two-dimensional shapes of the PN without taking into account the effects of projection. There have been several efforts to account for the variety of morphologies by different views of a single, unified, basic three-dimensional structure. Khromov and Kohoutek (1968) explain the morphology of PN in terms of a open ended cylinder projected onto the sky, and Masson (1989b, 1990), and Aaquist and Kwok (1996) employ an ellipsoidal shell model in which the PN morphology is determined by an ellipsoidal shell with both radial and angular density gradients ionized by a central star to different depths in different directions. Simulated images for 110 PN were produced by Zhang and Kwok (1998a), using the ellipsoidal shell model (see Fig. 8.8).

The situation we face today is not unlike the morphological classification of galaxies in the early part of the 20th Century. There have been two different approaches in interpreting the morphology of PN. One of them is the belief that the variety of morphologies represent different views of a single, unified, basic three-dimensional structure as envisioned by Khromov and Kohoutek (1968), whereas the other regards the range of shapes as an evolutionary sequence. Using CCD images of 51 PN taken in the light of low, moderate, and high ionization emission lines, Balick (1987) organized the nebulae into an evolutionary sequence consistent with the idea of the shaping of PN by interacting winds. The explanation of the diverse PN morphologies by interacting winds dynamics is given in Section 13.9.

9

Problems and questions

By the early 1970s, the field of PN had achieved a high degree of success. The nebular spectrum in the visible was reasonably well understood and PN had served well as a laboratory for atomic physics. Laboratory or theoretically derived atomic parameters such as recombination rates, collisional excitation rates, and spontaneous decay rates had been used to interpret the observed strengths of the line fluxes. The accounting of processes not observable in the terrestrial environment (e.g. the 2γ radiation, forbidden lines, etc.) is a particularly noteworthy accomplishment. The model of PN (which we refer to as the classical model), consisting of a nebular gas shell of fixed mass photoionized by a hot central star, seemed to be adequate in explaining the nebular spectrum. The combination of sophisticated observations (in particular spectroscopy) with theoretical calculations has made physics of gaseous nebulae one of the most successful examples of modern astrophysics.

Although astronomers were justifiably elated by the success of PN research, a number of problems were lurking under the surface. Here we summarize several examples of problems with the classical model that were starting to be recognized in the early 1970s.

- The nebular mass problem: in the classical model in which the PN is made up of a uniform-density shell of a fixed mass, the ionized masses of PN should be well determined by the measurement of the Hβ flux or the radio continuum flux (see Section 4.5). However, in cases in which the distances were reasonably well known, the actual derived masses were found to spread over several orders of magnitude, in contradiction to the traditional assumption of a fixed-mass nebula.

- The distance problem: the large observed range in nebular masses challenged the main assumption ($M_i \sim 0.2\ M_\odot$) used in the traditional method of distance determination: the Shklovsky method.

- The missing flux problem: although the nebular spectrum is often interpreted under Case B assuming that the nebula is ionization bounded, there were indications that at least some PN are not ionization bounded. If UV photons are leaking from the nebula, the total flux in the optical spectrum will not account for the total flux in the Lyman continuum. The strong infrared excesses found in PN also suggested that a significant amount of the nebula flux lies in the infrared, again pointing to the fact that the total emergent flux may have been underestimated.

- Placement of PN in the H-R diagram: because of the above two problems, the luminosities of PN were very poorly determined. The high temperature of the central star also necessitated an indirect method of temperature determination (see Section 7.1). This resulted in erroneous placements of PN in the H-R diagram and greatly impeded the development of a theoretical understanding of the evolution of the central star.
- The missing stellar mass problem: the detection of WDs in open clusters suggests that WDs descend from stars with initial masses as high as 8 $M_\odot$. However, the total observed nebular plus central star masses of PN is less than 1 $M_\odot$. Where has the remaining mass gone?
- The angular resolution problem: although hundreds of PN have been discovered by spectroscopic surveys, many were too small to be resolved by conventional optical telescopes. Most of the research on PN had concentrated on a few large and extended PN, and the properties of the stellar PN were poorly known.
- The evolution problem: the lifetime of the PN phenomenon can be estimated by their dynamical ages. Although we believe that this dynamical age should be comparable to the evolutionary age of the central stars, there were no quantitative models on how fast the central stars were evolving. In fact, the classical model assumes that the central star is static and emits a constant amount of ionizing photons. There was no understanding of how the central star evolves from a red giant to a CSPN with a over ten-fold increase in effective temperature.
- The dynamics problem: whereas a number of theories were proposed on how the atmosphere of a red giant could be detached under certain instabilities, none of them was quantitatively successful in ejecting the right amount of mass. The theoretical understanding of nebular expansion also remained primitive.
- The morphology problem: what is the origin of the diverse morphologies observed in PN? PN possess definite symmetries in their appearances, but they come in a variety of shapes.

The subsequent revision of the classical model and theories that revolutionizes our understanding of PN can be traced to a better understanding of the progenitors of PN: the asymptotic giant branch stars. Progress made in the understanding of the structure and circumstellar environment of AGB stars is primarily responsible for an improved understanding of the evolution of PN central stars and the origin of the nebula.

10

Asymptotic giant branch stars – progenitors of planetary nebulae

Stars are classified as low, intermediate, or high mass according to the nuclear reactions they undergo. Low-mass stars are defined as those that develop electron-degenerate He cores on the red giant branch (RGB). If the He core grows to $\sim 0.45\ M_\odot$, the star will undergo a core He flash until degeneracy is removed and quiescent He burning begins. Intermediate-mass stars can initiate core He burning under nondegenerate conditions and develop an electron-degenerate carbon-oxygen (C-O) core after core He exhaustion. At the completion of core He burning, low-mass stars also develop an electron-degenerate C-O core, and their subsequent evolution is similar to that of intermediate-mass stars. This is the beginning of the asymptotic giant branch. Stars that are massive enough can undergo He-shell flashes (also called thermal pulses) on the AGB. The AGB is terminated by either (a) complete removal of the hydrogen envelope by mass loss; or (b) ignition of carbon in the degenerate core. Massive stars are defined as those that develop a nondegenerate C-O core and therefore can ignite carbon nonviolently. They are able to go through a series of nuclear burnings (C, O, Ne, etc.), leading to the construction of the iron core followed by core collapse and supernova explosion.

The end products of evolution for low-, intermediate-, and high-mass stars are very different. For very-low-mass stars that have masses $<0.5\ M_\odot$ on the horizontal branch, they are unable to ascend to the AGB but evolve directly from the horizontal branch to become C-O WDs and gradually contract and fade away when H in the envelope is exhausted. The low-mass stars that are massive enough to evolve through the AGB will become C-O WDs after their H envelope has been lost in the PN phase. For intermediate-mass stars, if their core mass grow to exceed the Chandrasekhar limit (see Section 15.1), carbon detonation will occur under degenerate conditions and they will explode as supernovae. If the H envelope is completely depleted by mass loss when the core mass is still below the Chandrasekhar limit, they will end their evolution as C-O WDs. The evolution of massive stars ends with core collapse and supernova explosion and leaves behind a neutron star or a black hole.[†]

The upper initial mass limit for low mass stars is estimated to range from 1.8–2.2 $M_\odot$, depending on chemical composition and the treatment of convective boundaries (Iben

---

[†] Very low mass stars ($M_{MS} \leq 0.5\ M_\odot$) that cannot ignite He under degenerate conditions will evolve into He WDs. However, given the age of the Galaxy is smaller than the MS lifetime of such stars, the observed He WDs are likely to be the result of close binary evolution involving higher-mass stars.

and Renzini 1984).[‡] The minimum mass for massive stars is estimated to be between 7 and 10 $M_\odot$, or as low as 5–6 $M_\odot$ depending on overshoot, which is the dividing line between stars that will evolve to become WDs and neutron stars (or between PN and supernovae). The most critical test of this number is through the observations of WDs in open clusters. Since the ages of open clusters are well determined, the presence of a WD in a young cluster (such as the Pleiades) implies that the WD must have evolved from a star massive enough to complete its evolution within the cluster age. This has put the upper mass limit for WD formation (M_{WD}) to be $>8\ M_\odot$. (Romanishin and Angel, 1980; Reimers and Koester, 1982; Koester and Reimers, 1996). The high value of the WD mass limit suggests that mass loss on the AGB must be so efficient that carbon detonation under degenerate conditions never happens.

10.1   Structure of AGB stars

Low- and intermediate-mass stars develop an electron-degenerate C-O core after the exhaustion of helium in the core. This represents the beginning of the AGB phase of evolution. The term asymptotic is used because the $T_* - L_*$ relationship of low-mass AGB stars overlaps with the $T_* - L_*$ relationship of stars on the RGB. The AGB can be divided into two phases: (i) the early-AGB, where the H-burning shell extinguishes as the result of envelope expansion, and the luminosity is provided by He shell burning; and (ii) the thermal-pulsing AGB (TP-AGB), which begins with the reignition of H in a thin shell. Depending on the core mass, the lifetime of the early-AGB phase is $\sim 10^7$ yr (Iben and Renzini, 1983). During 90% of the TP-AGB, H-shell burning is the dominant source of energy. However, as the mass of the He shell below the H-burning shell increases, it will lead to an increase in the triple-α reaction rate and eventual thermonuclear runaway (Schwarzschild and Härm, 1965). The event, known as a He-shell flash or thermal pulse, lasts until the He shell overexpands and cools and the star returns to a steady state of H-shell burning. During the AGB, a star can go through a number of thermal pulses depending on its mass. During each of these episodes, several interesting events occur: (i) at the He flash, the temperature inside a flash-driven convective shell is high enough for N to be converted into Ne; (ii) after the flash, the base of the convective envelope reaches down past the H/He discontinuity and a dredge up of heavy elements occurs; (iii) for higher masses, nuclear burning can occur at the bottom of the convective envelope where C is converted to N and the stellar luminosity can reach very high values (Iben, 1991). The thermal pulse phase parameter (ϕ) is defined as zero at the H-shell flash, and unity at the beginning of the next flash. At $\phi \sim 0$, He is converted to C and O in the shell while the H shell is idle. As the He burning diminishes, dredge up occurs. A quiescent He-burning phase ($\phi < 0.15$) is followed by a quiescent H-burning phase. The contribution to the total luminosity by H and He burning as a function of ϕ is shown in Fig. 10.1.

The luminosity of an AGB star depends only on the mass of the core (M_c) and not on its total (core plus envelope) mass. For TP-AGB stars, the luminosity averaged over the

---

[‡] The possibility that convective elements can penetrate beyond the mathematical boundary of zero-buoyancy acceleration is referred to as "convective overshoot."

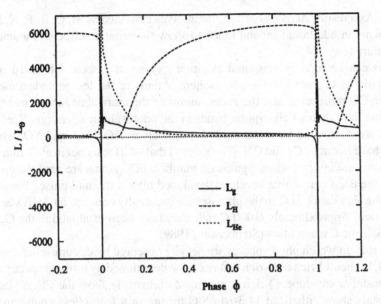

Fig. 10.1. The contribution of hydrogen burning (L_H), helium burning (L_{He}), and gravothermal energy release (L_g) to the surface luminosity as function of the thermal-pulse phase ϕ (from Schönberner, 1997).

He-flash cycle can be approximated by

$$L_* = 59,250(M_c/M_\odot - 0.52)L_\odot, \tag{10.1}$$

which is known as the core-mass luminosity relationship (Paczyński, 1970). If we take the maximum core mass to be the Chandrasekhar limit (see Section 15.1), then the maximum luminosity for an AGB star is $\sim 5 \times 10^4 \, L_\odot$. However, in high ($>5 \, M_\odot$) mass stars that develop deep convective envelopes with very high base temperatures ("hot bottom"), this limit can be exceeded (Blöcker and Schönberner, 1991; Boothroyd and Sackmann, 1992).

The time interval between the shell flashes can also be approximated by

$$\log \Delta t(\mathrm{yr}) = 7.47 - 0.49\frac{Z}{Z_\odot} + \left(-3.98 + 0.52\frac{Z}{Z_\odot}\right)\frac{M_c}{M_\odot} \tag{10.2}$$

(Vassiliadis and Wood, 1994), where Z is the metallicity. For $Z/Z_\odot = 1$, a star with $M_c = 0.6 \, M_\odot$ has an interpulse period of $\sim 8 \times 10^4$ yr. This decreases to $\sim 3 \times 10^3$ yr when $M_c = 1 \, M_\odot$.

10.2 Photospheric composition of AGB stars

Stars on the MS can be fitted into a simple spectral classification system (O, B, A, F, G, K, M) based on the strengths of the absorption lines of He, H, and metals, where each spectral class can be shown to have a one-to-one correspondence with effective temperature. However, the AGB on the H-R diagram is not just a mass sequence, but a superposition of mass and evolutionary sequences. Furthermore, because of the dredge up from the core, the surface composition undergoes significant changes over the lifetime of the AGB. As a result, AGB stars found themselves classified as M, C, R, S, N, J, and so on, and with an additional second index to show the separate effects of abundance and temperature (e.g., C4,3e).

AGB stars can be broadly classified as either oxygen or carbon rich based on the abundance ratio of C and O in the photosphere. Atoms of the less abundant element are tied up in CO molecules, and the excess atoms of the more abundant element form molecules that have distinct absorption bands in the photospheric spectrum. Spectra of O-rich stars are dominated by metal oxide bands such as TiO, ZrO, and YO, whereas C-rich stars have bands of C_2 and CN. It is believed that AGB stars begin as O-rich stars, and C-rich stars are created when significant numbers of C atoms are dredged up from the core by the deep convective envelope developed after a thermal pulse. The steady increase of the abundance of C in the photosphere eventually converts an AGB star from O rich to C rich. Approximately 6,000 C-rich stars have been cataloged in the General catalog of Galactic Carbon Stars (Stephenson, 1989).

For AGB stars in which photospheres are heavily obscured by circumstellar dust (see Section 10.4.1), their O-rich or C-rich nature can be determined by infrared spectroscopy of the circumstellar envelope. O-rich stars almost exclusively show the silicate feature, and C-rich stars show either the 11.3-μm SiC feature or a featureless continuum due to graphite or amorphous carbon. For stars that have both infrared and optical spectral classifications, the correlation between the infrared (circumstellar) and optical (photo-spheric) is excellent (Kwok *et al.*, 1997). Since the optical classification scheme stops at spectral type M10 because of dust obscuration, an infrared classification scheme allows an extension of the spectral classifications to more evolved AGB stars.

10.3   Pulsation on the AGB

Because of the large luminosity and radius, the envelopes of AGB stars are often unstable and undergo large-amplitude pulsations. These are known as long-period variables, or Mira variables (named after the first known variable star Mira Ceti, also known as o Ceti). Mira variables are radial pulsators driven mainly by H ionization effects and have typical pulsation periods of 200–600 days. They are found to obey a period-luminosity relationship, with more luminous Miras having longer periods. Such relationships are, however, more likely to be the result of mass rather than evolution-ary effects. Whether Miras pulsate in the fundamental mode or in a first overtone has been controversial (Wood, 1981; Willson, 1981). Nevertheless, the existence of a period-luminosity relationship suggests that Miras with $P < 400$ days have the same pulsation mode. Long-term monitoring of the infrared and OH light curves of infrared stars suggests that there exist AGB stars with periods much longer than those of optical Miras (see Section 10.4.2).

10.4   Mass loss on the AGB

Although mass loss from red giant stars has been known for a long time (Deutsch, 1956), the mass-loss rates can only be measured optically for a small number of bright stars, and the rates were not thought to be high enough to affect the stars' evolution. The advent of infrared and millimeter-wave astronomy in the early 1970s has made possible the direct observation of the ejected material, and mass-loss rates of 10^{-5} $M_\odot$ yr^{-1} were found to be common in late AGB stars. This implies that over a short period of $\sim 10^5$ yr, a significant fraction of the original mass of the star can be lost. Depending on the mass of the core, such mass loss rates can exceed the nuclear burning rate by 2 to 3 orders of magnitude (see Section 10.6). The lifetime and the termination of the AGB are therefore controlled by stellar winds from the surface but not nuclear burning in the core.

10.4.1 Infrared observations of mass loss

The first evidence of large-scale mass loss from AGB stars emerged as the result of the development of mid-infrared detectors. A number of stars were found by photometric observations to be much brighter in the infrared than in the visible, and to have very low (~ 600 K) color temperatures that could not be due to photospheric emission. The most likely interpretation is that these stars are surrounded by thick dust shells which absorb the visible light from the stars and reemit it in the infrared. In 1969, Woolf and Ney discovered a well-defined emission feature at the wavelength of 10 μm in a number of cool stars. Further broadband photometry (Gehrz and Woolf, 1971) revealed that this feature is common to most giants with spectral type later than M3. Woolf and Ney (1969) attributed this feature to silicate grains. Chemical equilibrium calculations by Gilman (1969) supported this identification by showing that silicate grains should be the first solid particles to condense in O-rich stars. For carbon stars, the first condensates should be carbon and silicon carbide (SiC). The 11.3-μm SiC feature was subsequently detected by Treffers and Cohen (1974) in carbon stars.

Although early observations of infrared excesses were limited to known optical M stars, the *Air Force Geophysical Laboratory (AFGL)* rocket survey of the sky at 10 and 20 μm discovered that many sources with large infrared excesses had no optical counterparts. Ground-based follow-up observations showed that many of these sources are late-type stars with thick circumstellar envelopes that completely obscure the central star. Because of the high opacity of the grains, they experience significant radiation pressure and cannot remain stationary with respect to the star. Thus, there is no doubt that the infrared excess is a clear manifestation of mass loss.

Spectrophotometric observations of Mira variables and infrared stars discovered in the *Two Micron Sky Survey (IRC)* and the *AFGL* surveys show that the silicate feature goes from emission to deep absorption, which is the result of an increasing optical thickness in the silicate band (see Fig. 10.2). The fact that objects with silicate absorption features often have very red colors supports the hypothesis that these objects are undergoing mass loss at a high rate (Volk and Kwok, 1988).

10.4.2 Circumstellar OH maser emission

Mass loss on the AGB is also manifested in molecular emission. Many O-rich AGB stars are found to have OH maser emission (Herman and Habing, 1985a). The OH emission often shows a two-peaked profile, which is a manifestation of a steady mass-loss process with the expansion velocity of the envelope being given by half of the velocity separation of the two peaks (Kwok, 1976).

Many infrared sources discovered in the *IRC* and *AFGL* surveys are found to be OH sources. Such OH sources with no optical counterparts are known as OH/IR stars. Since

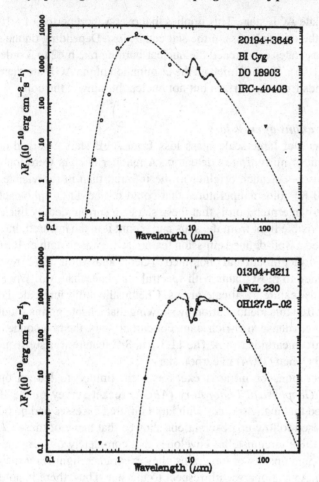

Fig. 10.2. The SED of two oxygen-rich AGB stars.

the publication of the *IRAS Point Source Catalog* (PSC), searches of *IRAS* sources with colors similar to known OH/IR stars have yielded many more OH sources (Eder *et al.*, 1988; Lewis *et al.*, 1990; te Lintel Hekkert *et al.*, 1991). The total number of OH/IR stars cataloged is now over 2000.

The correlation of variations between the OH and infrared fluxes suggests that the OH maser is pumped by infrared photons. Monitoring of the variability of a group of ~50 unidentified OH/IR stars shows a mean period of 1,000 days, much greater than the mean period of 350 days for optical Mira variables (Herman and Habing, 1985a). Since only ~30% of optical Mira Variables have OH emission (Bowers, 1985), it is likely that OH emission only develops when the mass-loss rate is high enough to create a saturated maser, and OH/IR stars probably represent stars further up the AGB than optical Mira variables.

Figure 10.3 shows the *IRAS* color-color diagram of three groups of O-rich AGB stars: (i) stars optically classified as spectral type M with the 9.7 μm silicate feature in emission

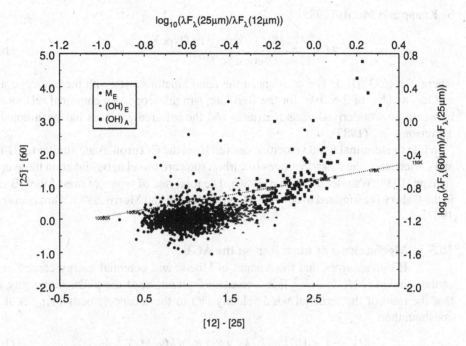

Fig. 10.3. IRAS color-color diagram of O-rich AGB stars.

(M_E); (ii) stars with no optical spectral classification with OH emission showing the silicate feature in emission (OH_E); and (iii) stars with no optical spectral classification with OH emission showing the silicate feature in absorption (OH_A). This figure clearly shows that the three groups of O-rich stars form a continuous color sequence with decreasing color temperatures (Kwok, 1990). This provides strong evidence that the OH/IR stars represent an extension of the AGB formed by M-type stars. Beyond the spectral type M10, circumstellar extinction of the photosphere becomes too large for optical spectral classification, and circumstellar spectroscopy (both infrared and radio) is the only way to identify these stars.

10.4.3 Thermal CO emission

Thermal CO emission from AGB stars was first detected by Solomon *et al.* (1971). The observed ^{12}CO and ^{13}CO profiles in the carbon star CW Leo (IRC+10216) show characteristic flat-topped and double-peaked profiles, which were respectively interpreted by Kuiper *et al.* (1976; see Section 5.3) as the optically thick and optically thin cases of the thermally excited molecular spectra in a partially resolved circumstellar envelope created by continuous mass loss. Subsequent CO surveys have resulted in the detection of several hundred AGB stars in the $J = 1 - 0, 2 - 1$, and $3 - 2$ transitions (Knapp *et al.*, 1982; Loup *et al.*, 1993; Neri *et al.*, 1998). A semitheoretical formula relating the mass-loss rate to the observed CO $J = 1 - 0$ antenna temperature is given by Knapp and Morris (1985):

$$\dot{M} = \frac{(T_A/K)(V/\mathrm{km\ s^{-1}})^2(D/\mathrm{pc})^2}{\mathrm{const} \times f^{0.85}} M_\odot\ \mathrm{yr}^{-1},\qquad(10.3)$$

where $f = [CO]/[H_2]$. The constant in the denominator varies with the telescope used and has a value of 2×10^{15} for the Bell Lab 7-m telescope. An empirical relationship between the CO-derived mass loss rates and the infrared excesses has been noted by Thronson *et al.* (1987).

While the terminal wind velocities derived from the CO profiles are similar for O-rich and C-rich stars, the mass-loss rates of carbon stars are found to be higher on the average (Knapp, 1987; Wannier and Sahai, 1986). The total rate of return of mass by AGB stars in the Galaxy is estimated to be $\sim 0.3\ M_\odot\ \mathrm{yr}^{-1}$ (Knapp and Morris, 1985; Thronson *et al.*, 1987).

10.5   Mechanisms of mass loss on the AGB

If one assumes that the amount of kinetic and potential energy carried in the stellar wind $[1/2\dot{M}(V^2 + V_{esc}^2)]$ is a constant fraction, α, of the stellar luminosity, and that the ratio of the terminal wind velocity (V) to the escape velocity (V_{esp}) is also a constant, then

$$\dot{M}_R = 4 \times 10^{-13}\eta(L/L_\odot)(R_*/R_\odot)(M_*/M_\odot)^{-1}\ M_\odot\mathrm{yr}^{-1},\qquad(10.4)$$

where

$$\eta = 8 \times 10^4 \frac{\alpha}{(1 + V^2/V_{esc}^2)}\qquad(10.5)$$

(Willson, 1987). Equation (10.4) was first empirically obtained by Reimers (1975) to approximate stellar mass loss, including the solar wind. The Reimers formula is simple, and the underlying assumption of its origin is independent of the actual mass-loss mechanism and therefore has been widely used in the literature as a working formula for mass loss. One could, of course, attempt to improve on it by introducing physics into the formula provided that there is an adequate knowledge of the mass-loss mechanism.

Mass loss mechanisms for AGB stars can be generally classified into (i) thermally driven, (ii) wave-driven, and (iii) radiation pressure-driven models. Discussions on these mechanisms can be found in the reviews of Cassinelli (1979), Castor (1981), and Holzer and MacGregor (1985). Since stars with high mass-loss rates all show large infrared excesses (see Section 10.4.1), it is likely that grains play a role in the mass-loss mechanism. Because of the large opacity of the dust grains to stellar radiation, there is no doubt that the grains will be ejected by radiation pressure. The only remaining question is whether the dust can effectively carry the gas with it. The observed optical depths at 10 μm of circumstellar envelopes of AGB stars reach as high as 50 (Volk and Kwok, 1988), suggesting that the grains are not ejected at high terminal velocities of thousands of kilometers per second as they would be if unhindered. The grain and gas components are in fact tightly momentum coupled as the result of grain-gas and gas-gas collisions (Gilman, 1972), and radiation pressure on grains will dominate the dynamics of the circumstellar envelope once they form (Kwok, 1975). There have been questions raised on whether dust can form close enough to the photosphere to actually cause the mass loss.

One possible solution is to appeal to hybrid models in which shock waves created by pulsations in the atmosphere raise the atmospheric scale height and, therefore, the gas density at the dust condensation point (Bowen, 1988).

Mass loss formulae for AGB evolution can be divided into two categories: (i) a steady increase in mass-loss rate throughout the AGB; or (ii) a low (Reimers type) mass-loss rate during most of the AGB and a dramatic increase near the end of the AGB (a "superwind"). An example of the latter is the mass loss formula of Baud and Habing (1983):

$$\dot{M}_{BH} = 4 \times 10^{-13} \left(\frac{M_{en,0}}{M_{en}} \right) \frac{(L_*/L_\odot)(R_*/R_\odot)}{(M_*/M_\odot)} M_\odot \text{yr}^{-1} \tag{10.6}$$

where $M_{en,0}$ is the envelope mass at the beginning of the OH-emitting phase. Since the envelope mass (M_{en}) approaches zero at the end of the AGB, the mass-loss rate increases sharply near the end.

The existence of an infrared sequence in the *IRAS* color-color diagram of AGB stars with silicate features (see Fig. 10.3) suggests that the mass-loss rate is steadily increasing on the AGB. A steady mass loss formula was introduced by Volk and Kwok (1988) to fit this infrared sequence:

$$\dot{M}_{VK} = 1.8 \times 10^{-13} \left[\frac{M_*(0)}{8 M_\odot} \right] \frac{(L_*/L_\odot)(R_*/R_\odot)}{(M_*/M_\odot)} M_\odot \text{ yr}^{-1}, \tag{10.7}$$

where $M_*(0)$ is the total mass of the star at the beginning of the AGB. This formula was improved upon by Bryan *et al.* (1990) to produce better agreements with the observed initial mass-final mass relationship (see Section 10.7):

$$\dot{M}_{BVK} = 1.15 \times 10^{-13} \left\{ \left[\frac{M(0)}{M_\odot} \right]^2 - 10.6 \left[\frac{M(0)}{M_\odot} \right] + 10.2 \right\}$$

$$\times \frac{(L_*/L_\odot)(R_*/R_\odot)}{(M_*/M_\odot)} M_\odot \text{yr}^{-1}. \tag{10.8}$$

Vassiliadis and Wood (1993) assume that the mass-loss rate is a power law of pulsation period (P):

$$\log \dot{M}(M_\odot \text{ yr}^{-1}) = -11.4 + 0.0123 \, P(\text{days}) \tag{10.9}$$

until it reaches the maximum value of

$$\dot{M} = \frac{L_*}{cV}. \tag{10.10}$$

Blöcker (1995a) assumes that mass loss occurs at the Reimers rate ($\eta = 1$) until the pulsation period exceeds 100 days. After this point, the mass-loss formula becomes:

$$\dot{M}_B = 4.83 \times 10^{-9} (M_*/M_\odot)^{-2.1} (L_*/L_\odot)^{2.7} \dot{M}_R \tag{10.11}$$

The decreasing total mass and increasing luminosity with time cause the mass-loss rate to increase steadily on the AGB.

The mass-loss formulae in Eqs. (10.6)–(10.11) reflect much better the high mass-loss rates deduced from infrared and molecular line observations. They give good approximations to the ending of the AGB by a complete removal of the H envelope through mass loss. In addition, these empirical formulae provide a necessary element in the computation of AGB evolution.

The use of the term superwind is unfortunate because it implies a qualitative change in the mass-loss mechanism and a sudden change in the mass-loss rate. The term has its origin in comparison with the Reimers formula, which has no physical basis and is totally inappropriate for AGB mass loss. The use of the Reimers formula is popular only because it is simple and convenient. However, as measured mass-loss rates from infrared and millimeter observations show, the mass-loss rate throughout the entire AGB is high and the distinction between ordinary winds and superwinds is therefore irrelevant.

10.6   AGB evolution with mass loss

As the result of H shell burning, the rate of increase of the core mass is given by

$$\dot{M}_c = \frac{L_*}{x E_H},\qquad(10.12)$$

where $x = 0.7$ is the H-mass abundance and $E_H = 6 \times 10^{18}$ erg is the energy released by converting 1 g of H to He. The He created by H burning will be converted to C by the triple-α process during the thermal pulse episodes. Combining Eqs. (10.1) and (10.12) we have

$$\frac{dM_c}{dt} = 8.2 \times 10^{-7} \left(\frac{M_c}{M_\odot} - 0.5 \right) M_\odot \ \text{yr}^{-1}.\qquad(10.13)$$

The rate that the H envelope is consumed by nuclear burning is therefore significantly smaller than the observed mass-loss rates of $10^{-6} - 10^{-4}$ $M_\odot$ yr^{-1}. Since the core and envelope of an AGB star are effectively decoupled, the ascent of the AGB can be approximated by the increasing mass of the core that is due to nuclear burning and the decreasing mass in the envelope that is due to mass loss from the surface. Integrating Eq. (10.13), we have

$$\frac{M_c}{M_\odot} = A e^{t/t_0} + 0.5,\qquad(10.14)$$

where $t_0 = 1.2 \times 10^6$ yr. The zero point in time ($t = 0$) can be defined to begin with the first thermal pulse, and mass loss can be assumed to commence at the same time. The mass of the core at $t = 0$ [$M_c(0)$] has been calculated for low-mass (1–3 $M_\odot$) stars of solar abundance by Boothroyd and Sackmann (1988b). The corresponding values for higher mass (3–9 $M_\odot$) stars are given by Iben and Renzini (1983) and Blöcker (1995a). The following empirical formulae are adopted by Bryan *et al.* (1990):

$$\frac{M_c(0)}{M_\odot} = 0.483 + 0.021 \frac{M_*(0)}{M_\odot}, \qquad\qquad 1.25 < \frac{M_*(0)}{M_\odot} < 3,$$

$$= 0.546 + 0.087 \left\{ \left[\frac{M_*(0)}{3 M_\odot} \right]^2 - 1 \right\}, \qquad 3 < \frac{M_*(0)}{M_\odot} < 9.\qquad(10.15)$$

Although Eq. (10.14) cannot replace a detailed numerical stellar structure model, it does give an approximate description of AGB evolution. Together with the core-mass luminosity relationship, Eq. (10.14) suggests that an AGB evolves with increasing

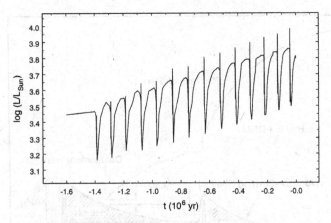

Fig. 10.4. Time evolution of the stellar luminosity of a 3 $M_\odot$ star on the AGB. The spikes represent thermal pulses. The luminosity gradually increases over the lifetime of the AGB as the core mass increases.

luminosity at an approximately exponential rate. The mass-loss rate, if also dependent on luminosity, will also increase as the star ascends the AGB. The end of the AGB is therefore a race between nuclear burning in the core and mass loss from the surface.

As we have seen, mass loss plays a significant role in the AGB evolution, and, yet until the 1980s, no numerical stellar evolution models had included mass loss. Schönberner (1979, 1981, 1983) was the first to take mass loss into account, although the Reimers formula he used greatly underestimated the actual mass-loss rate on the AGB. Later model calculations by Boothroyd and Sackmann (1988a–1988d) also used Reimers rates only. More realistic mass loss formulae were used in the models of Vassiliadis and Wood (1993) and Blöcker (1995a).

Figure 10.4 shows the evolution of a 3 $M_\odot$ star ending its AGB evolution with a core mass of 0.605 $M_\odot$ from the model of Blöcker (1995a). We can see that after each thermal pulse, there is a brief (~500 yr) pulse of surface luminosity followed by a substantial luminosity dip, which lasts for ~20 – 30% of the pulse cycle. The luminosity then gradually increases with time until the next thermal pulse.

The evolution of the structure of a TP-AGB star is illustrated in the schematic diagram shown in Fig. 10.5. The changes in the masses of the C-O core, He- and H-burning shells, and the H envelopes are plotted as a function of time. The discrete jumps in the masses occur at thermal pulses. The mass of the H envelope decreases with time as the result of mass loss from the surface. Between the thermal pulses, H burning steadily increases the mass in the He shell. At each thermal pulse, He burning decreases the mass of the He shell and adds to the mass of the core (lower solid curve in Fig. 10.5). The cross-hatched areas refer to the convective zones formed during the He flash and a dredge-up episode mixes He, C, O, and Ne into the surface. This cycle repeats until the H envelope is completely depleted by mass loss.

10.7 Initial mass-final mass relationship

The first indication that the upper mass limit for WD formation ($M_{\rm WD}$) can be as high as 6 $M_\odot$ came from the discovery of a WD in the Pleiades (Luyten and Herbig,

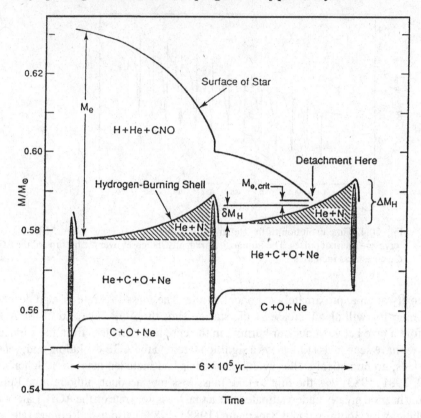

Fig. 10.5. Schematic of the structure of a TP-AGB star (from Iben, 1995).

1960). This is derived by postulating that the WD progenitor mass must be larger than the cluster turnoff mass. Romanishin and Angel (1980), who found faint blue WD candidates in southern clusters by searching deep red and blue photographic plates, estimated that $M_{WD} \sim 7 \, M_\odot$. Reimers and Koester (1982) and Koester and Reimers (1996) found four high-mass WD ($M > 0.9 \, M_\odot$) in NGC 2516 and estimated that $M_{WD} \simeq 8 \, M_\odot$. The large difference between the initial mass (M_i, mass on the MS) and final mass provide the indirect proof that significant mass loss has occurred on the AGB.

Masses of WDs can be determined in the following way. A spectrophotometric analysis with model atmospheres allows the determination of the surface gravity, from which mass can be determined by using a mass-radius relationship. If the cooling age can also be determined (see Section 15.2), then the initial mass of the WD progenitor can be estimated by equating the cluster age minus the cooling age to the theoretical evolutionary age from the MS to the AGB. Using this method, Weidemann and Koester (1983) found the initial mass-final mass relationship to be flat in the range $M_i = 1 - 3 \, M_\odot$ and then bend upward sharply to $M_{WD} \sim 8 \, M_\odot$.

A number of approximate formulae have been proposed for the initial mass-final mass relationship. For example, the following formula is used by Vassiliadis and Wood (1994):

$$M_c/M_\odot = 0.473 + 0.084(M_i/M_\odot) - 0.058 \log(Z/Z_\odot). \qquad (10.16)$$

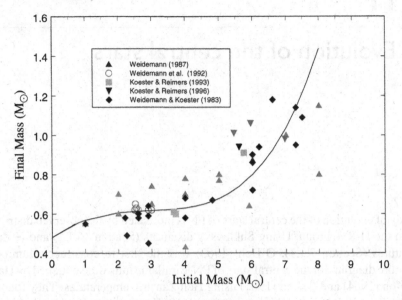

Fig. 10.6. The initial mass – final mass relationship.

In Fig. 10.6, we show the observed distribution of initial and final masses and a fit with a fourth order polynomial. This curve is used to calculate the theoretical distribution of PN in Section 18.2.

10.8   Summary

The advent of infrared astronomy has led to the discovery of many evolved AGB stars, both O and C rich, that possess thick circumstellar envelopes and whose photospheres are completely obscured by circumstellar dust. These stars represent an extension of the AGB, previously known to be occupied by only optical Mira variables. Their extensive circumstellar envelopes are built up by a large-scale mass loss process, driven by a combination of pulsation and radiation pressure. The mass-loss rates measured by infrared and millimeter-wave techniques are much higher than the nuclear burning rates, and mass loss, not nuclear burning, is responsible for the termination of the AGB. Mass loss will continue until the H envelope is completely removed, and the effective temperature of the exposed core increases to such an extent that the mass loss mechanism is no longer operable. This sets the stage for post-AGB evolution toward the PN phase.

The AGB phase, which lasts only ~10^6 yr, totally changes the eventual fate of a star. Mass loss from the surface prevents the core from growing to a size needed to ignite carbon, and therefore it allows stars with initial masses as high as 8 $M_\odot$ to become WDs. The mass loss process also builds up a considerable circumstellar envelope (in both mass and size) around the star. These materials will eventually supply the mass necessary to build a PN.

11

Evolution of the central stars

The study of evolution of the central stars of PN is motivated by the observed distribution of PN in the H-R diagram. Using Shklovsky distances (Section 16.1.1) and H Zanstra temperatures (Section 7.1.1), O'Dell (1963) was the first to construct a luminosity-temperature diagram for the central stars of PN. Similar results were obtained by Harman and Seaton (1964) and Seaton (1966) using HeII Zanstra temperatures. They found that both low ($T_* = 30,000$ K) and high temperature (10^5 K) central stars had low luminosities (10^2 L$_\odot$) and that the intermediate temperature (50,000 K) stars have high luminosities (10^4 L$_\odot$). Apparently, the distribution of PN in the H-R diagram had the shape of an upside-down horse shoe, which was interpreted as an evolutionary sequence and named the Harman-Seaton sequence.

Harman and Seaton (1964) suggested that PN were formed at the end of the horizontal branch, followed by a rapid increase in temperature and luminosity of the central star. Such rapid evolution in 10^4 yr (the dynamical age of PN) posed a great challenge to theorists. Early efforts have concentrated on remnant stellar cores without nuclear burning, and undergoing gravitational and thermal adjustments while contracting toward the WD stages. Although artificial models with specific initial conditions could be made to mimic the Harman-Seaton sequence, the results are far from satisfactory (cf. Salpeter, 1968; Shaviv, 1978). Deinzer (1967) started with the right assumption of having an electron-degenerate core with a thin H envelope, but he had to modify his models in order to fit the Harman-Seaton sequence (Deinzer and von Sengbusch, 1970).

Unfortunately, much of these theoretical efforts were wasted because the observed PN distribution in the H-R diagram was in error. In order to plot a star on the H-R diagram, one needs to know the luminosity and temperature. For most PN, the central star is not directly observable, and its temperature has to be inferred from nebular properties. The problems and the associated errors of such exercises are discussed in Section 7.1.1. In order to obtain the luminosity, one needs to know the distance and the total emitted flux. Distance determination has proved to be very difficult for PN, and the often-used Shklovsky method is not reliable (see Section 16.1.1). Modern observations have shown that PN emit a significant fraction of their total flux outside of the visible, so the observed fluxes were often under-estimated in the past. The uncertainty of whether all PN are ionization bounded also compounded the missing flux problem.

In view of the misleading observational data, it is particularly remarkable that Paczyński (1971a) came up with the correct evolutionary tracks based on theoretical considerations alone. This is one of the rare cases in the history of astronomy in which theory leads the progress of the field and in fact led to the revision of the observational data that first appeared to have contradicted the theory.

11.1 The modern era

Our modern understanding of central star evolution began with the models of Paczyński (1971a). He obtained post-AGB evolutionary tracks by beginning with AGB stars with H envelopes added on cores of various masses. Since the luminosity of a red supergiant is only dependent on its core mass but not its envelope mass [see Eq. 10.1], he argued that the progenitors of PN must have similar luminosities as the central stars of PN. Since central stars of PN have luminosities of $\sim 10^4$ L$_\odot$, the only candidates for the progenitors of PN are double-shell-burning AGB stars. Since the effective temperature of an AGB star changes very little as long as the star has an extensive H envelope, he was able to show that an AGB star will not evolve to the blue side of the H-R diagram until the H envelope mass (M_e) had dropped to a very small value. For a star of core mass of 0.6 $M_\odot$ to evolve to an effective temperature of 30,000 K (just hot enough for photoionization of the nebula), he estimated that M_e must be less than 4×10^{-4} $M_\odot$. For a core mass of 1.2 $M_\odot$, the corresponding value for the H envelope mass was 10^{-6} $M_\odot$.

These findings had profound implications on our understanding of PN evolution. Such small H envelope masses means that they will be rapidly consumed by H burning, and the lifespan of PN must be very short. Over such short time scales, the core masses are essentially unchanged and the central stars of PN will evolve with constant luminosity until H is exhausted. This is in contradiction with the observational data at the time, which suggested that the luminosities of central stars should increase with temperature (along the Harman-Seaton sequence) before turning down to the cooling track.

The evolution of the central stars of PN can be separated into two phases:

   (i) when M_e is approximately a few percent of the total mass, the envelope begins to shrink, but it is still able to release enough gravitational energy to maintain the temperature needed for nuclear burning at the base of the envelope. The luminosity of the star stays approximately constant, and the star evolves horizontally across the H-R diagram.

   (ii) When H burning depletes M_e to $<10^{-4}$ $M_\odot$, the H-burning shell starts to die out, and the luminosity drops; the star moves to the cooling track toward the WD phase.

The time scale for the crossing of the H-R diagram (Δt) is determined by the total amount of available fuel (i.e. M_e) and the fuel consumption rate:

$$\Delta t = \frac{M_e}{\dot{M}_c},$$

(11.1)

where $\dot{M}_c$ is the nuclear burning rate. For a core mass (M_c) of 0.6 $M_\odot$, $\dot{M}_c$ is $\sim 10^{-7}$ $M_\odot$ yr^{-1}. If M_e is $\sim 10^{-3}$ $M_\odot$, then Δt is $\sim 10^4$ yr. However, Δt is a very strong function of the core mass (Renzini, 1982). According to an approximate formula given by Tylenda (1989),

$$\log \Delta t \sim 7.3 - 5.9(M_c/M_\odot) \quad M_c > 0.7 \, M_\odot,$$

(11.2)

so a star of 1.0 $M_\odot$ will have a post-AGB nuclear burning lifetime of only 25 yr! The fading time (t_f) of PN from $T_* = 30,000$ K to the point where the stellar luminosity has dropped by a factor of 10 from the horizontal track is also highly dependent on the core mass. For H-burning models,

$$t_f \sim 1.6 \times 10^4 (0.6\ M_\odot/M_c)^{10.0} \tag{11.3}$$

(Iben, 1984). For central stars in the core mass range of $0.56 - 0.64\ M_\odot$, $t_f \sim 32,000 - 8,000$ yr. Since the observed dynamical ages of PN have similar values, the observed population of PN can be well explained by central stars in this mass range undergoing H burning (Schönberner and Weidemann, 1983).

11.2   Evolutionary models with mass loss and thermal pulses

The Paczyński models were improved by Schönberner (1979, 1981), Kovetz and Harpaz (1981), Iben (1984), Wood and Faulkner (1986), Vassiliadis and Wood (1994), and Blöcker (1995b) by including the effects of thermal pulses. The Schönberner models show that H-burning central stars evolve approximately three times faster than He-burning stars in the parts of the H-R diagram where PN are found. Given the constraints imposed by dynamical time scales, observed PN can be very well explained by H-burning models with core masses between 0.55 and 0.64 $M_\odot$. If this is the case, then one can conclude that PN ejection is not initiated by a thermal pulse but occurs during the quiescent H-burning phase on the AGB (Schönberner, 1987). The Schönberner tracks and the change in M_e for stars with various M_c are shown in Figs. 11.1 and 11.2, respectively.

If one defines the end of the AGB as the time when the large-scale mass loss has stopped, the residual H envelope mass at that temperature may still be high enough that the movement on the H-R diagram can be very slow. Table 11.1 shows the evolutionary

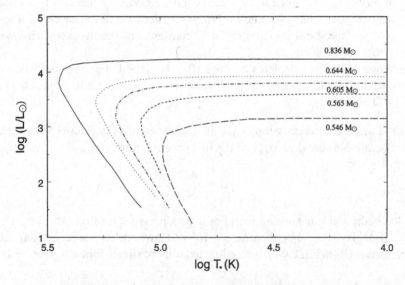

Fig. 11.1. The Schönberner H-burning evolutionary tracks for stars with core masses from 0.546 to 0.836 $M_\odot$.

Table 11.1. *Evolutionary times for central stars of PN*

M_c ($M_\odot$)	$\dot{M}_c$ ($M_\odot$ yr^{-1})	Δt (yr) log T_* (K)			
		3.5–3.7	3.7–4.0	4.0–4.5	4.5–5.0
1.2	6.0×10^{-7}	4×10^5	6000	3	0.7
0.8	2.4×10^{-7}	4×10^5	5300	100	120
0.6	9.0×10^{-8}	5×10^5	5300	1700	3500
0.57	6.0×10^{-8}	5×10^5	7000	6000	18000

Table taken from Schönberner (1987).

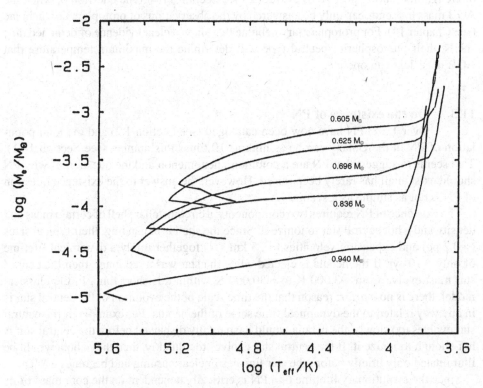

Fig. 11.2. The H envelope mass as a function of the effective temperature for several H-burning models of PN evolution (from Blöcker, 1995b).

rates in different parts of the H-R diagram for central stars of different core masses. One can see that all models evolve very slowly at $T_* < 5,000$ K. For the lower mass models ($M_c < 0.6\ M_\odot$), the times needed to reach 30,000 K (the beginning of the PN phase) are much higher than the dynamical ages of PN. This problem can be avoided by introducing post-AGB mass loss. By application of a Reimers type of mass loss formula in the post-AGB phase, the mass-loss rate exceeds the nuclear burning rate until $T_* \sim 10^4$ K. This will have the effect of reducing the transition time to $\sim 10^3$ yr (Schönberner, 1983).

Blöcker (1995b) assumes that the end of AGB mass loss is related to the pulsational properties of the star. As the star begins to evolve blueward as the result of a thinning H envelope, the period of radial pulsation decreases. As the period decreases from 150 to 50 days, pulsationally-driven mass loss is no longer feasible, and this is defined as the end of the AGB (corresponding to T_* between 6,000 and 6,500 K). Beyond this point, the post-AGB mass loss is again assumed to be of the Reimers type until the central star is hot enough that radiation pressure on resonance lines takes over as the mass-loss mechanism. The mass-loss rate in this regime is assumed to be $\propto L^{1.9}$ as found by Pauldrach et al. (1988).

The choice of the termination point of AGB mass loss has significant implications on the subsequent evolution. Vassiliadis and Wood (1994) assume an earlier termination of mass loss, which results in longer transition times to PN and more frequent occurrence of He ignition during post-AGB evolution (see Section 11.4). The question of when the AGB mass loss ends can only be answered by the observations of protoplanetary nebulae (see Chapter 14). For protoplanetary nebulae that show a clear evidence of detached dust shells, their photospheric spectral type will determine the maximum temperature that AGB mass loss can operate.

11.3   On the existence of PN

Over 1,500 PN have now been cataloged (see Section 1.2) and the total population of PN in the Galaxy can be as high as 10 times this number (see Section 18.1). This seems to suggest that PN are a common phenomenon and the question of why PN should exist at all has rarely been raised. However, the answer to the existence question of PN is not as simple as it seems.

The existence of PN requires two components: a circumstellar shell of certain mass and density and a hot central star to ionize it. Since the shell is expanding, their typical sizes ($\sim$0.2 pc) and expansion velocities ($\sim$25 km s^{-1}) together imply a dynamical lifetime of only $\sim$10^4 yr. If the nebula is ejected when the star was a red giant, then the central star must evolve from $\sim$3,000 K to $\sim$30,000 K within this short time. In the classical model, there is no *a priori* reason that the time scale of the evolution of the central star is in any way related to the dynamical time scale of the nebula. For example, if the central star evolves too slowly, the nebula would have totally dispersed before the central star is hot enough to ionize it. If the central star evolves too quickly, then the nebula would be illuminated only briefly before the star finishes nuclear burning and becomes a WD.

Since the evolutionary lifetime of a PN is critically dependent on the core mass of its central star, on one hand a low-mass star will evolve too slowly to become PN (termed "lazy PN" by Renzini, 1982). On the other hand, a high-mass ($>1\ M_\odot$) star will be luminous too briefly to be detected. Consequently, only stars with core masses in a very narrow mass range ($\sim$0.6 $M_\odot$) will be seen as PN. The actual transition time (and therefore the existence of PN) is critically dependent on the mass of the H envelope at the end of AGB. This suggests that the evolution of the central star is closely tied to the ejection mechanism. If the AGB is ended by a sudden ejection, there will always be an undetermined amount of mass left around the core. This will result in a long delay in the evolution to high temperature and no PN will form. However, if the H envelope is depleted by a steady stellar wind, the wind will continue until the stellar temperature has changed significantly. This can be the result of the H envelope being two small for pulsation to occur, or it can be because the effective temperature has increased to such an extent that dust can no longer form close to the photosphere. In either case, a steady mass loss scenario will ensure that the minimum amount of mass remains in the envelope and the least amount of transition time between AGB and PN phases (Kwok, 1982a).

11.4   Helium-burning central stars

If a star leaves the AGB with a thermal pulse cycle phase of $0 \leq \phi \leq 0.15$, the post-AGB evolution of the star will be dominated by He burning. In the H-R diagram, their tracks resemble the quiescent H-burning tracks but their evolutionary speed is slower. If a star leaves the AGB at $\phi \gtrsim 0.9$, a last thermal pulse can occur during the post-AGB evolution, changing the star from H burning to He burning. When this occurs, the outer edge of the convective zone (see Fig. 10.5) in the He-burning region extends into the H envelope (Iben, 1984) and the star will evolve back to the AGB and restart the post-AGB evolution as a He burner after a short stay of a few hundred years.

Figure 11.3 shows the post-AGB evolution of a star with a core mass of 0.625 $M_\odot$ when it leaves the AGB at $\phi = 0.87$. At first, the star evolves horizontally like the normal H burners. When the thermal pulse occurs, the luminosity drops by a factor of 10 in less 10 yr and the star evolves back to the AGB. After reaching a luminosity maximum, the star evolves to the blue again as a He burner (born again), but at a much reduced speed. Because of post-AGB mass loss and slow evolution, the H envelope is

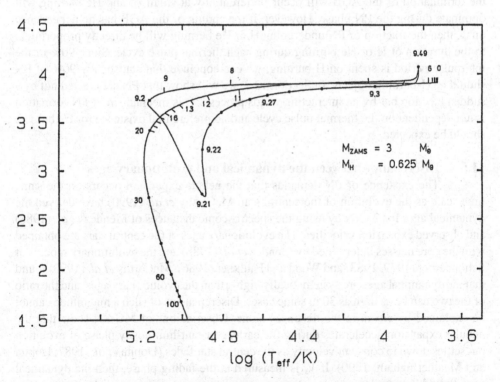

Fig. 11.3. Post-AGB evolution model for a star with a core mass of 0.625 $M_\odot$ that leaves the AGB at $\phi = 0.87$. The tick marks show time in 10^3 yr (from Blöcker, 1995b).

much reduced in mass and H is only reignited mildly. He burning continues to provide most of the luminosity for the rest of the evolution until the WD cooling stage is reached (see Section 15.2).

One feature of this evolutionary scenario is that the CSPN go through two separate phases of high luminosity. The possibility exists that the nebulae can be excited again by UV photons during the second phase and be observed as a large, low-surface-brightness PN. The PN A30 and A78 have been suggested to be in such phase of evolution.

Theoretically, if the interpulse period is short enough, a second thermal pulse can occur during the post-AGB evolution. This will require a long transition time, or a large remnant mass. Given the fact that the interpulse periods for stars with core masses <1 $M_\odot$ are $>10^4$ yr [Eq. (10.2)], repeated shell flashes must be very rare because the dynamical lifetime of PN is not much longer.

It would be interesting to relate the observed H-deficient central stars (see Section 7.2.1) to the He-burning episodes of post-AGB evolution. In practice, however, since the He-burning tracks cover a wide area of the H-R diagram, it is difficult to assign an individual star to a specific track. From the observed ratio between H-rich and He-deficient stars in the MC PN sample, Dopita *et al.* (1996) estimated that $\sim$45–70% of the central stars are He burners. This is in contradiction to the conclusions drawn from galactic PN.

He-burning models have strong appeal in particular to people who associate the PN ejection with a thermal pulse event (see Chapter 12). If the PN ejection is brought about by a sudden event caused by the luminosity peak during a thermal pulse, then the termination of the AGB will occur preferentially at small ϕ, and He burning will dominate during the PN phase. However, if the ending of the AGB has no dependence on ϕ, then the fraction of PN undergoing H or He burning will be directly proportional to the duration of H or He burning during each thermal pulse cycle. Since 90% of the interpulse period is spent on H burning, we can conclude that statistically 90% of PN should be H-burning objects. In Section 12.2, we will show that PN are created not by a sudden ejection but by an interacting winds process. This mechanism of PN formation is not dependent on the thermal pulse cycle and no preferential existence for He burners should be expected.

11.5 Discrepancy between the dynamical and evolutionary ages

The existence of PN demands that the nebula expansion occurs on the same time scale as the evolution of the central star. McCarthy *et al.* (1990) have derived the dynamical ages for 23 PN by using the spectroscopic distances of Méndez *et al.* (1988) and observed expansion velocities. The evolutionary ages of the central stars are obtained with the core masses determined by Méndez *et al.* (1988) and the evolutionary models of Schönberner (1979, 1983) and Wood and Faulkner (1986). McCarthy *et al.* (1990) found that the dynamical ages are systematically higher than the evolutionary ages, and the ratio of the two can be as high as 30 in some cases. Discrepancies of such a magnitude cannot be explained by errors in the distances alone. Observations of MC PN show that the nebular expansion accelerates during the early, constant-luminosity phase of evolution but settles down to constant velocity as the central star fades (Dopita *et al.*, 1987; Dopita and Meatheringham, 1990). If V_s is measured at the fading phase, then the dynamical time should be even larger.

One major source of uncertainty in the evolutionary time is the zero point of the evolutionary tracks. Since the evolutionary age is highly sensitive to the residual mass of the H envelope above the core, a small increase in the assumed value of the residual mass can greatly lengthen the evolutionary age. A smaller mass-loss rate than the Reimers rate used in Schönberner (1983) will also increase the evolutionary time.

If evolutionary age is not underestimated, can the dynamical be over-estimated? One possibility suggested by McCarthy *et al.* (1990) to explain the discrepancy is that the use of the outer radii of PN as their size will lead to the overestimation of dynamical age. The real question is, is the material represented by the outer radius expanding at the observed expansion velocity? The expansion velocity (V_s) is traditionally determined from the peak separations of the line of a particular ion. However, because of the ionization structure of PN, different ions will yield different expansion velocities (Chu *et al.*, 1984; Weinberger, 1989). Line profiles measure the velocities of regions with the highest emission measure, and those velocities can be much smaller than the velocity as defined by the expansion of the Hβ image. Both the density and velocity structures are functions of both radius and age. The angular size is often determined from a certain brightness cutoff (which is proportional to the minimum density measurable), while the expansion velocity is often measured from regions of maximum density. It is therefore questionable that the dynamical time can be obtained from the simple expression R/V_s. These questions are only resolved when detailed models of dynamical evolution of PN are available (see Section 13.5).

11.6 Summary

We now believe that PN descend from double-shell-burning AGB stars. The central stars are made of the electron-degenerate C-O cores of their progenitors, and they are burning H (in most cases) in a thin envelope. Because of the low value of M_e during PN evolution, CSPN evolve across the H-R diagram at constant luminosity before turning downward to the cooling tracks when H is exhausted. The maximum luminosity and temperature of CSPN are strictly functions of the core mass. In order to synchronize its evolution across the H-R diagram with the expansion of the nebula, the existence of PN demands that the core masses of central stars of PN be in a narrow range ($0.55-0.64\,M_\odot$).

Some aspects of the evolutionary model remain uncertain. Post-AGB mass loss was introduced to shorten the transition time between the AGB and PN in order to bring the evolutionary time more in line with the observed dynamical time. The Reimers type of mass loss was used only as a matter of convenience, for details of the post-AGB mass loss process are still poorly known. The observation of a post-AGB wind, and the possible identification of a physical mechanism, have to wait until the discovery and observation of protoplanetary nebulae, objects in transition between the AGB and PN (see Chapter 14).

The existence of H-deficient central stars remains a mystery. If they are He-burning objects, then the fact that He burners evolve slower than H burners suggests that their numbers should be higher than the strict 10% fraction expected from the theory of thermal pulses.

Most interestingly, the central-star evolutionary theories require a small residual H envelope mass at the end of the AGB in order to obtain a reasonable transition time. This has serious implications on the dynamical theories of PN ejection.

12

Formation of planetary nebulae

Traditional theories of PN ejection appeal to a variety of sudden ejection mechanisms including dynamical instabilities induced by recombination (Roxburgh, 1967; Lucy, 1967), pulsational instabilities (Kutter and Sparks, 1974; Wood, 1974; Tuchman *et al.*, 1979), envelope relaxation oscillations due to thermal instability in the core (Smith and Rose, 1972), radiation pressure (Faulkner, 1970; Finzi and Wolf, 1971), or thermal pulses (Härm and Schwarzschild, 1975; Trimble and Sackman, 1978). The pulsation models represent an extension of the theory of Mira pulsation and predict that finite amplitude pulsation in the fundamental mode is not possible, leading to subsequent relaxation oscillations and PN ejection (Wood, 1981). The thermal-pulse theories rely on the luminosity peak just after the helium flash, which could lead to greater pulsational instability as the result of an enlarged stellar radius. Although these sudden-ejection models are intuitively appealing, none of them is quantitatively successful in ejecting the right amount of mass.

The difficulties of the sudden-ejection models should have been apparent given Paczyński's results. Paczyński's models have shown that the central star will remain at low effective temperatures if the envelope mass is higher than $\sim 10^{-3} M_\odot$. It is impossible for any instability mechanism to be so precise in leaving behind just the right amount of mass. For example, if the sudden ejection removes $0.199 M_\odot$ instead of $0.2 M_\odot$, the star will stay on the red side of the H-R diagram long after the (never-ionized) nebula is completely dissipated.

Even if PN are created by a sudden impulsive ejection, an additional pressure is needed to maintain the expansion of the nebula and to prevent it from falling back onto the central star. Various mechanisms such as pressure from the ionization front (Capriotti, 1973; Wentzel, 1976) and radiation pressure on grains (Ferch and Salpeter, 1975) have been considered but none was found to be adequate. Mathews (1966) used gas pressure as the driving force for expansion and included photoelectric heating and [OIII] cooling in the energy equation. He found that the nebula will disperse very quickly, and suggested that continuous mass loss is needed to avoid backfill.

The discovery of mass loss from AGB stars (Section 10.4) created additional problems for the dynamical models of PN. PN are not expanding into a vacuum as previously believed, and the effects of a pre-existing circumstellar envelope on the dynamics of PN expansion have to be considered.

12.1 Effects of AGB mass loss on the formation of planetary nebulae

The discovery at the University of Minnesota of extensive mass loss on the AGB by infrared (Gehrz and Woolf, 1971) and millimeter-wave (Solomon *et al.*, 1971) observations had significant impact on our understanding of the origin of PN. Given the high mass-loss rates ($\sim 10^{-5}$ $M_\odot$ yr^{-1}) inferred by these observations, it occurred to this author that over the lifetime of an AGB star ($\sim 10^6$ yr), as high as several solar masses of material can be lost. Furthermore, 10^{-1} $M_\odot$ of stellar wind material had just been ejected in the past $\sim 10^4$ yr, which is comparable to the dynamical lifetime of PN. Since the mass in PN is only ~ 0.1 $M_\odot$, the remnant of AGB mass loss must have an effect on the formation of PN. The possibility that the circumstellar envelopes of the AGB progenitor could be responsible for the formation of PN was also considered by Paczyński in a short paper that he published in *Astrophysics Letters* (Paczyński, 1971b). However, this simple model has several difficulties:

(i) The observed expansion velocities of PN are higher than the stellar wind velocities of AGB stars;
(ii) The observed densities of PN shells are higher than the densities in the AGB envelopes;
(iii) Many PN have well-defined shell structures with sharp inner and outer boundaries, whereas the circumstellar envelopes of AGB stars have smooth, diffuse structures.

PN are therefore not simply AGB envelopes diffusing into the interstellar medium; a separate mechanism is needed to accelerate, compress, and shape the AGB envelope into PN (Kwok, 1981a, 1982).

12.2 The interacting stellar winds model

The solution to these problems was found in the Interacting Stellar Winds (ISW) model of Kwok *et al.* (1978). They suggest that the PN phenomenon does not represent a separate ejection of matter, but just a rearrangement of material ejected over a long period of time. Assuming AGB mass loss can continue until most of the H envelope has been depleted and the core exposed, nuclear burning will gradually remove the remaining thin H envelope and the star will evolve to the blue. When the central star is hot enough, a new phase of mass loss is initiated under the mechanism of radiation pressure on resonance lines. Since the radius of the star is now much smaller than when it was on the AGB, the terminal velocity of the wind (which is related to the escape velocity) will be much higher. This new high-speed wind will run into the remnant of the AGB wind and sweep up the old wind like a snow plough and create a high-density shell. Since the swept-up shell is compressed on both sides by dynamical pressure, it will develop a definite shell structure as observed in PN.

The ISW model was first formulated for the momentum-conserving case. Assuming that the $t = 0$ corresponds to the time that the slow wind ends and the fast wind begins, and the shell is formed by an inelastic collision between the two winds, the mass of the shell (M_s) at time t is

$$M_s = \int_{Vt}^{R_s} \frac{\dot{M}}{V} dr + \int_{R_s}^{vt} \frac{\dot{m}}{v} dr,$$

$$= \left(\frac{\dot{M}}{V} - \frac{\dot{M}}{v} \right) R_s - (\dot{M} - \dot{m})t, \tag{12.1}$$

where R_s is the shell radius at time t, $\dot{M}$ and V are the mass-loss rate and expansion velocity of the AGB wind, and $\dot{m}$ and v are the mass-loss rate and expansion velocity of the central-star wind. The stellar radius is assumed to be much smaller than Vt and vt. From the conservation of momentum, the equation of motion can be written as

$$M_s \ddot{R}_s = \frac{\dot{m}}{v}(v - \dot{R}_s)^2 - \frac{\dot{M}}{V}(V - \dot{R}_s)^2. \tag{12.2}$$

Equation (12.2) can be integrated to give

$$R_s = \frac{\dot{M} - \dot{m}}{\frac{\dot{M}}{V} - \frac{\dot{m}}{v}} t + \sqrt{\frac{(\zeta t + B)^2 - A}{\zeta}}, \tag{12.3}$$

where

$$\zeta = \frac{\dot{M}\dot{m}}{Vv} \frac{(v - V)^2}{(\dot{M}/V - \dot{m}/v)^2} \tag{12.4}$$

and A and B are constants to be determined from boundary conditions. At large t, Eq. (12.3) reduces to the terminal velocity solution given by setting the left hand side of Eq. (12.2) to zero

$$\dot{R}_s = \frac{(\dot{M} - \dot{m}) + (v - V)\sqrt{\dot{M}\dot{m}/vV}}{\dot{M}/V - \dot{m}/v}. \tag{12.5}$$

Assuming $\dot{M} = 10^{-5} M_\odot$ yr^{-1}, $V = 10$ km s^{-1}, $\dot{m} = 10^{-8} M_\odot$ yr^{-1}, and $v = 2{,}000$ km s^{-1}, we find that Eqs. (12.5) and (12.1) give $V_s = \dot{R}_s = 14$ km s^{-1} and $M_s = 0.05 M_\odot$ after 10^4 yr. Although these values are comparable to the observed expansion velocities and masses of many PN, there are PN with higher expansion velocities or masses and it is likely that not all PN can be explained by the ISW model under the momentum-conserving case.

The momentum-conserving case assumes that all the excess energy of the fast wind is radiated away. Since the observed CSPN wind speeds are much higher than the sound speed, the shocks generated can create very high temperature gas in the postshock region if cooling is not efficient. If part of the energy of the fast wind is transformed into thermal energy, then thermal pressure will provide additional acceleration to the nebular shell. The ISW model was extended to the energy-conserving case by Kwok (1982), Kwok (1983), Kahn (1983), and Volk and Kwok (1985). Since there are three mass components (the slow wind, the fast wind, and the swept-up shell) in the system, two shock fronts will be generated. An outer shock propagates into the slow wind and the inner shock toward the star. Because of the high velocity of the central-star wind, the Mach number of the

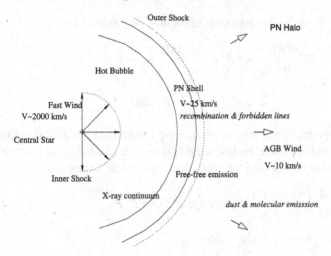

Fig. 12.1. Schematic diagram of the ISW model.

inner shock is very high and the postshock region will have a very high temperature. As a result the inner shock is likely to be adiabatic because of the low efficiency in radiative cooling at high temperatures. This high-temperature region is referred to as the bubble.

In contrast, the relative velocity between the AGB wind and the shell is moderate, and the temperature in the postshock region is likely to be low enough to allow for the existence of metal ions for forbidden-line cooling. Therefore the outer shock can be treated as isothermal. Since the density jump across an isothermal shock is proportional to the square of the Mach number and is not limited to a maximum value of 4 as in an adiabatic shock, the shell density can be very high and the thickness of the shell will be thin. A schematic diagram of the ISW model is shown in Fig. 12.1.

In order to simplify the problem, a contact discontinuity is assumed to separate the hot bubble and the swept-up shell across which no gas or heat passes. The gas pressure is assumed to be unchanged across the contact discontinuity but the temperature and density take on discontinuous values. Thus the shell is made up of swept-up slow wind only, and gas in the hot bubble is all from the central-star wind. From the conservation of mass, the rate that mass is being swept up is given by

$$\frac{dM_s}{dt} = 4\pi r^2 \rho(r)(\dot{R}_s - V),$$ (12.6)

where $\rho(r) = \dot{M}/(4\pi r^2 V)$ is the density profile of the AGB wind. The shell mass at time t is therefore given by

$$M_s = \frac{\dot{M}}{V}(R_s - Vt).$$ (12.7)

At a given time, the swept-up shell is pushed by the thermal pressure (P) of the hot bubble and impeded by the remnant of the AGB wind. Since the momentum flux from the AGB wind is $\frac{dM_s}{dt}(\dot{R}_s - V)$, the equation of motion for the shell is

$$M_s \frac{d^2 R_s}{dt^2} = 4\pi R_s^2 P - \left[\frac{dM_s}{dt}(\dot{R}_s - V)\right]$$

$$= 4\pi R_s^2 P - \frac{\dot{M}}{V}(\dot{R}_s - V)^2. \qquad (12.8)$$

Assuming no energy exchange (the adiabatic approximation), we find that the total energy input into the bubble from the fast wind ($\frac{1}{2}\dot{m}v^2$) must be balanced by change in the internal energy in the bubble and the work done because of expansion:

$$\frac{d}{dt}\left[\frac{4\pi}{3}R_s^3\left(\frac{3}{2}P\right)\right] = \frac{1}{2}\dot{m}v^2 - 4\pi R_s^2 P \frac{dR_s}{dt}. \qquad (12.9)$$

From Eq. (12.8), the pressure in the bubble is

$$P = \frac{1}{8\pi R_s^2}\frac{\dot{M}}{V}\frac{d^2}{dt^2}(R_s - Vt)^2 \qquad (12.10)$$

and Eq. (12.9) can be written as

$$\frac{1}{R_s^2}\frac{d}{dt}\left[R_s^3\left(2\pi R_s^2 P\right)\right] = \frac{\dot{m}v^2}{2}. \qquad (12.11)$$

Substituting Eq. (12.10) into Eq. (12.11), we have

$$\frac{\dot{M}}{4R_s^2 V}\frac{d}{dt}\left[R_s^3\frac{d^2}{dt^2}(R_s - Vt)^2\right] = \frac{\dot{m}v^2}{2}. \qquad (12.12)$$

This inhomogeneous nonlinear differential equation can be solved numerically to give R_s as a function of time. However, if we are only interested in the steady-state solution, a similarity solution can be found by assuming that $R_s \propto t^\alpha$. By requiring all terms to have the same t dependence, we find $\alpha = 1$ and

$$R_s = V_s t, \qquad (12.13)$$

where $V_s = \dot{R}_s = $ constant is the solution to the following cubic algebraic equation:

$$\left(\frac{\dot{M}}{V}\right)V_s^3 - 2\dot{M}V_s^2 + \dot{M}VV_s = \frac{1}{3}\dot{m}v^2. \qquad (12.14)$$

Substituting Eq. (12.13) to Eqs. (12.7) and (12.8), we have

$$M_s = \dot{M}\left(\frac{V_s}{V} - 1\right)t, \qquad (12.15)$$

$$P = \frac{\frac{1}{2}\dot{m}v^2}{6\pi V_s^3}t^{-2}. \qquad (12.16)$$

Using typical observed values for the two winds, that is, $\dot{M} \sim 10^{-5}~M_\odot$ yr^{-1}, $V \sim$ 10 km s^{-1}, $\dot{m} \sim 10^{-8}~M_\odot$ yr^{-1}, and $v \sim 2{,}000$ km s^{-1}, we have $V_s \sim 30$ km s^{-1} and $M_s \sim 0.15~M_\odot$ after 10^4 yr. These values are higher than the corresponding values found in the momentum conserving case, demonstrating that the ISW model has no difficulty explaining the observed properties of PN.

Since the sound speed in the hot bubble is high, the pressure everywhere in the bubble will be quickly equalized and we can assume that the shocked fast wind is isobaric. The average density in the bubble is

$$\rho_{bubble} = \frac{\dot{m}t}{\frac{4}{3}\pi(V_s t)^3}. \tag{12.17}$$

Applying the pressure jump condition across the inner shock, we have

$$P = \frac{3}{4}\rho_1 v^2, \tag{12.18}$$

where

$$\rho_1 = \frac{\dot{m}}{4\pi r^2 v} \tag{12.19}$$

is the density of the unshocked wind in front of the inner shock. The location of the inner shock (R_1) can be found by substituting Eq. (12.16) into Eq. (12.18):

$$R_1 = \sqrt{\frac{9}{4}\left(\frac{V_s}{v}\right)}\, R_s. \tag{12.20}$$

For $v = 2,000$ km s^{-1} and $V_s = 30$ km s^{-1}, $R_1 \sim 0.17\, R_s$, implying that 99% of the volume interior to the shell is shocked. Combining Eqs. (12.16) and (12.17) and using the ideal gas law as the equation of state, we find that the temperature in the hot bubble is

$$T_{bubble} = \frac{\mu' m_H v^2}{9k}, \tag{12.21}$$

where μ' is the mean atomic weight per particle. For $\mu' = 0.6$ and $v = 2.000$ km s^{-1}, the temperature of the shocked gas is 3×10^7 K.

The efficiency of the interacting winds mechanism can be determined by comparing the ratios of the energy and momentum of the shell to those of the central-star wind. The energy efficiency ε can be defined as

$$\varepsilon = \frac{\frac{1}{2}M_s V_s^2}{\frac{1}{2}\dot{m}v^2 t}. \tag{12.22}$$

Assuming that the mechanical energy of the central-star wind greatly exceeds that of the AGB wind, we can reduce Eq. (12.14) to

$$V_s = \left(\frac{\dot{m}v^2 V}{3\dot{M}}\right)^{\frac{1}{3}}. \tag{12.23}$$

Substituting Eqs. (12.15) and (12.23) into Eq. (12.22), we have

$$\varepsilon \sim \frac{1}{3}. \tag{12.24}$$

The momentum efficiency can be defined as

$$\Pi = \frac{M_s V_s}{\dot{m} v t}.$$ (12.25)

Substituting Eqs. (12.15) and (12.23) into Eq. (12.25), we have

$$\Pi \simeq \frac{1}{3^{\frac{2}{3}}} \left(\frac{\dot{M} v}{\dot{m} V} \right)^{\frac{1}{3}},$$ (12.26)

which has a value of 28 using the wind parameters above and $\Pi = 19$ using the exact solution of Eq. (12.14).

It has been argued that the ISW process is inadequate to drive the expansion of PN because the observed momentum of PN shells exceeds that of the momentum in the central-star winds. The above analysis shows that in the adiabatic approximation, the theoretical momentum efficiency is $\sim$1,900% and has no problem accounting for the observations. A similar reasoning can also be used to resolve the "momentum paradox" found by millimeter-wave observers in bipolar outflows from star-formation regions (Kwok and Volk, 1985; Dyson 1984).

12.3   Transition from the momentum-conserving case to the energy-conserving case

At the early phase of interacting winds, the inner shock is located close to the shell and the dynamical evolution will begin in the momentum-conserving mode as cooling is effective. As the PN evolves the dynamics will change from momentum conserving and R_1 will gradually moves toward the star. The transition to energy conserving occurs when the cooling time (t_c) exceeds the time (t_{fill}) needed to fill the volume inside R_s with shocked gas. Assuming $R_1 = R_s$,

$$t_{\text{fill}} = \frac{\frac{4}{3} \pi \rho R_s^3}{\dot{M}}.$$ (12.27)

Using the jump conditions for a strong adiabatic shock,

$$\rho_{\text{bubble}} = 4\rho_1$$

$$= 4 \left(\frac{\dot{M}}{4\pi R_s^2 v} \right).$$ (12.28)

The cooling time can be expressed as

$$t_c = \frac{K^{3/2}}{q}$$ (12.29)

(Kahn, 1976) where $K = P/\rho_{\text{bubble}}^{5/3}$ and $q = 4 \times 10^{32}$ cm^6 g^{-1} s^{-1}. Combining Eqs. (12.18), 12.28), (12.29), and (12.27), we can derive the transition time:

$$t_{\text{tr}} = 5.23 \frac{q\dot{m}}{V_s v^5}$$ (12.30)

Using the same wind parameters as before, we have $t_{tr} \sim 10^{-4}$ yr, so the transition is very fast if v is large.

12.4 Observational confirmations of the ISW model

Although the ISW model was successful in explaining many of the features of PN such as density, expansion velocity, and the shell structure, the model also makes several predictions. First, it predicts that there should be a faint halo as the result of the remnant of the AGB envelope outside of the high-density nebular shell. Secondly, high-speed winds should be common in PN central stars. Thirdly, thermal X-ray emission may arise from the high-temperature bubble.

Confirmations of the second prediction came quickly. The *IUE* satellite was launched in January 1978, and one of the first scientific results from the *IUE* was the discovery of P-Cygni profiles suggestive of high-speed winds from the central stars of many PN (Heap *et al.*, 1978; see Section 7.5).

Although optical halos around PN have been known for a long time, it was not until the development of CCDs that halos were found to be a common occurrence (see Section 8.4). The use of CCD cameras have made possible the high-dynamic-range imaging of PN, which reveals the presence of faint halos outside of the bright PN shells. Spectroscopic observations of PN halos can directly determine the kinematics of the halo and compare the results to the predictions of the ISW model (Bässgen and Grewing, 1989).

Evidence for the presence of the remnant AGB envelope can also be found in the molecular envelopes in PN (see Section 5.5). The line profiles of the molecular lines in PN show a remarkable resemblance to the line profiles seen in AGB envelopes, suggesting that they share a common origin.

Because of the low emission measure in the hot bubble, X-ray emissions from PN are expected to be weak (Volk and Kwok, 1985). Extended X-ray emissions from PN were first reported by Kreysing *et al.* (1993), but all except one of the extended-emission nature were later found to be spurious (Chu *et al.*, 1993). Because of the limited spatial resolution and sensitivity of earlier X-ray telescopes, the only confirmed cases of diffuse X-ray emission are NGC 6543 (Kreysing *et al.*, 1993), Abell 30 (Chu *et al.*, 1997), and BD+30 3639 (Leahy *et al.*, 1998). However, X-ray emission in A30 comes predominantly from the fast stellar wind is ablating clumpy circumstellar material close to the star, and may not be caused by the hot bubble. Figure 12.2 shows the X-ray distribution over an optical image of BD+30 3639.

An alternate way to search for emission from the hot bubble is by spectral analysis. Leahy *et al.* (1994) found the PN NGC 7293 to have a two-component X-ray spectrum, where one component is of low temperature and is consistent with central star photospheric emission, and the second component is consistent with emission from a high-temperature, optically-thin plasma similar to that expected from the hot bubble (see Fig. 12.3). Also plotted in Fig. 12.3 are the ROSAT spectrum of NGC 6853, which has a pure photospheric spectrum, and the spectrum of BD+30 3639, which is consistent with emission from the hot bubble.

12.5 Summary

The ISW model casts completely new light on our understanding of the origin of the PN phenomenon. PN, which traditionally have been treated as an independent subject, are now related to the previous phase of stellar evolution, the AGB. The theories of stellar structure and mass loss, as well as the observational studies of the circumstellar

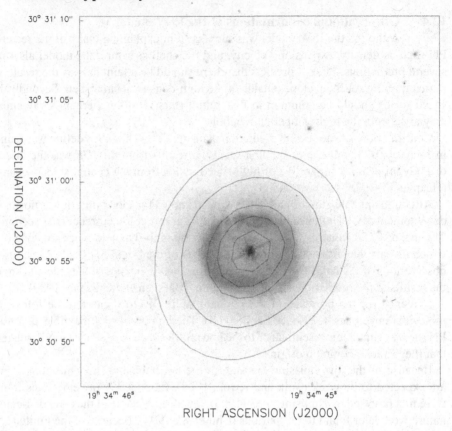

Fig. 12.2. The HST WFPC2 image of BD+30 3639 overlaid by ROSAT HRI X-ray contours.

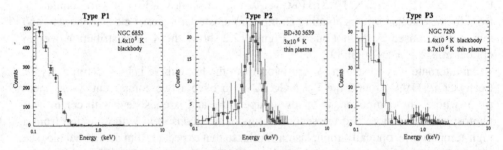

Fig. 12.3. Examples of ROSAT spectra of PN (from Conway and Chu, 1997).

envelope of AGB stars, are now relevant. The formation of PN is not produced by a sudden ejection as traditionally believed, but is the result of a new fast wind compressing and accelerating material ejected over the AGB lifetime. This mechanism also has implications on our understanding of the CSPN evolution. Since mass loss occurs throughout the AGB, it is unlikely that the depletion of the H envelope and the departure from the AGB occur exactly at $\phi = 0$. This implies that CSPN that begin their evolution as He burners must be extremely rare.

Most importantly, we now realize that the central star is not only radiatively interacting with the nebula but is also injecting mechanical energy into the nebula. Although the AGB progenitor provides the mass that makes up the nebula, the central star is responsible for feeding the energy required for the continued dynamical evolution of the nebula. This mechanical interaction leads to new radiation mechanisms for the nebula: thermal X-ray continuum radiation from the hot bubble and shock heating of molecular hydrogen (Section 5.5).

It is also worth mentioning that even though the ISW model was first formulated to understand the dynamical evolution of PN, it has been applied to a variety of other astrophysical phenomena, including ring nebulae around massive stars (WR stars, luminous blue variables, and blue supergiants), supernovae (Lou and McCray, 1991, Blondin and Lundqvist, 1993), young star outflows (Frank and Noriega-Crespo, 1994), relativistic jets in active galactic nuclei (Eulderink and Mellema, 1994), and even galactic superbubbles (MacLow *et al.*, 1989).

13

Dynamical evolution of planetary nebulae

We now recognize that a planetary nebula is a dynamical system whose nebular evolution is closely coupled to the evolution of the central star. The existence of a planetary nebula depends on the nebular and central star components evolving in step with each other. A complete description of the PN phenomenon therefore requires the following elements:

1. Evolution model of the central star [$L_*(t)$ and $T_*(t)$]. Other than the question of whether central stars of PN are predominately hydrogen or helium burning, one major uncertainty is the extent of mass loss in the post-AGB phase. Since the evolution time from the end of AGB to the beginning of photoionization is critical for the existence of PN, a better estimate on the mass-loss rate during the post-AGB phase is needed.

2. Winds from central stars of PN [$\dot{m}(t)$ and $v(t)$]. Whereas mass loss during the post-AGB phase affects the transition time to PN, mass loss during the PN phase has crucial effects on the dynamics of the nebula. Not only does the wind from the central star compress and accelerate the nebular shell, it also shapes the morphology of the nebula. On the observational side, the line profiles can be used to measure the terminal velocity and the mass-loss rate. Since the winds are likely to be driven by radiation pressure on resonance lines, theoretical estimates on $\dot{m}$ can also be made.

3. Ionization structure. Since the central star is evolving rapidly, the amount of Lyman continuum photons also changes significantly throughout the lifetime of the PN. During the early stages of evolution, most of the central star flux is absorbed by the dust component and reemitted as infrared radiation. As the central star temperature increases, an increasing amount of the central star flux goes into the gas component through the process of photoionization. When the CSPN enters the cooling track and its luminosity declines, the nebula may change from ionization bounded to density bounded, although observationally this transition is difficult to determine.

    As the ionization structure evolves with time, the atomic species available for cooling will also change. This can in turn lead to changes in the kinetic temperature of the gas, the sound speed, and the strengths of the shocks.

A complete model of PN must incorporate coupled, time-dependent solutions to the equations of stellar evolution, hydrodynamics, and line and continuum (dust) radiation

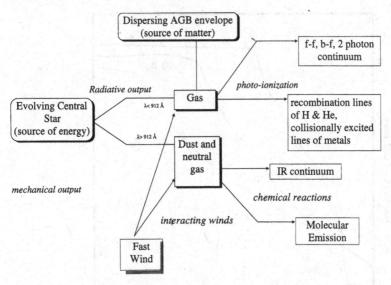

Fig. 13.1. Schematic showing the interactions between the central star and the nebula.

transfer (Fig. 13.1). Line profiles and images change greatly as the PN evolve. It is important to remember that the optical properties of PN (morphology, kinematics, etc.) reflect only the ionized parts of the nebula, and proper interpretation of the observed data is only possible if one keeps in mind the evolving nature of PN.

13.1 Variability of the central-star wind

As the central star evolves across the H-R diagram at a constant luminosity, its effective temperature changes by a factor of 10 and its radius contracts correspondingly by a factor of 100. Since the terminal velocity of the wind is related to the escape velocity of the star, the wind speed will not remain constant as the central star evolves. From the theory of radiation pressure on resonance lines as the mechanism of mass loss, Abbott (1978) gives an empirical formula

$$v = 3V_{esc}$$

$$= 3\sqrt{\frac{2GM_*}{R_*}}, \tag{13.1}$$

for winds from OB supergiants undergoing mass loss. Heap (1982) finds that the wind velocity is closer to four times the escape velocity based on *IUE* data. Observationally, central-star wind becomes undetectable when the surface gravity (g) reaches 10^5 cm s^{-2} (Cerruti-Sola and Perinotto, 1985). These values are reached at $\sim$2,800, 6,000, and 12,500 yr after the end of the AGB for the 0.64, 0.60, and 0.57 $M_\odot$ H-burning Schönberner models, respectively (Fig. 13.2).

The central-star mass-loss rate is also expected to vary with time. Schönberner (1983) assumes that the post-AGB mass loss is given by the Reimers formula [Eq. (10.4)].

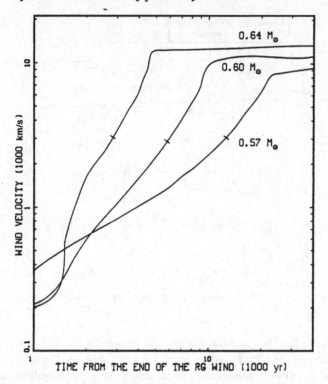

Fig. 13.2. Change of the central-star wind velocity as a function of time since the AGB. The three curves correspond to the 0.64, 0.60, and 0.57 $M_\odot$ models. Also marked on each curve is the point where the surface gravity is 10^5 cm s^{-2} (from Volk and Kwok, 1985)

Combining Eqs. (10.4) and (13.1), we find that the mechanical luminosity of the wind ($L_w = \frac{1}{2}\dot{m}v^2$) is directly proportional to L_*. Since L_* is constant over much of the PN lifetime, the constant L_w approximation used in Section 12.2 is not bad after all. In the radiation-driven wind model of Pauldrach *et al.* (1988), the central-wind mass-loss rate can be approximated by the following formula:

$$\log \dot{m} = -14.80 + 1.84 \, \log(L_*/L_\odot). \tag{13.2}$$

This formula is used by Marten and Schönberner (1991) in their PN dynamical evolution model. For the constant luminosity phase of a H-burning 0.6 $M_\odot$ star, $L_* = 6300 \, L_\odot$, and $\dot{m}$ has a value of $1.6 \times 10^{-8} \, M_\odot$ yr^{-1}, which is lower than the nuclear burning rate of $8.6 \times 10^{-8} \, M_\odot$ yr^{-1}.

If the CSPN wind velocity is low at the early phase of evolution, the nebula will take longer to make the transition from momentum conserving (Section 12.3).

The possibility that pressure from the ionized gas can accelerate the neutral part of the shell and induce Rayleigh-Taylor instabilities is investigated by Breitschwerdt and Kahn (1990). Later numerical models of Marten and Schönberner (1991), Frank (1994), and Mellema (1994) incorporate more accurate treatments of the gas radiation processes,

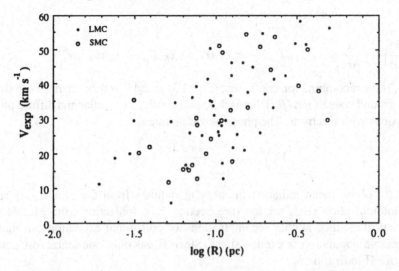

Fig. 13.3. Expansion velocity vs. radius plot for the Magellanic Cloud PN. The open and closed circles represent SMC and LMC PN respectively (from Dopita & Meatheringham, 1990).

the transition from momentum- to energy-driven flow, and the dynamical effects of ionization.

13.2 Nebular acceleration

In order to follow the dynamical evolution of PN, it would be useful to seek a correlation between the expansion velocities and the radii of PN. The $V_s - R_s$ relation is difficult to obtain for galactic PN because of distance uncertainties. However, for Magellanic Cloud PN, distance is not a problem. Figure 13.3 shows a V_s versus R_s plot for the LMC and SMC PN (Dopita and Meatheringham, 1990), and a weak correlation can be seen. As we have seen in Section 12.2, if L_w is a constant, V_s is a constant. If the nebular acceleration is real, then it implies that L_w is increasing with time (Volk and Kwok, 1985). For example, if $L_w \propto t^\gamma$ and the AGB wind has a density profile of $\rho \propto r^\beta$, the similarity solutions give $R_s \propto t^\alpha$, where

$$\alpha = \frac{\gamma + 3}{\beta + 5}. \tag{13.3}$$

(Dyson, 1981; Volk and Kwok, 1985). For $\beta = -2$ and $\gamma = 1$, $\alpha = 4/3$, or $V_s \propto t^{1/3} \propto R_s^{1/4}$. Unfortunately, the weak dependence of V_s on R_s does not provide a good test of the change of the post-AGB mass-loss formula.

13.3 Time-dependent ionization structure

Since the evolutionary time scales for high-mass central stars are very short, the static approximation for the ionization structure may not be applicable. If we define $x_{\beta,i}$ as the degree of ionization of ion β in ionization state i, then the ionization equation is given by

$$\frac{\partial x}{\partial t} + v\frac{\partial x}{\partial r} = x_{\beta,i-1}\Gamma_{\beta,i-1} - n_e x_{\beta,i}\alpha_A^{\beta,i} - x_{\beta,i}\Gamma_{\beta,i} + n_e x_{\beta,i+1}\alpha_A^{\beta,i+1} \qquad (13.4)$$

where α_A is the recombination coefficient [Eq. (2.31)] and Γ is the total rate of ionization from the ground state of ion (β,i) that is due to absorptions of stellar and diffuse photons and collisions with electrons. The photoionization rate is

$$\Gamma_{\beta,i}^{photo} = \int_{\nu_{\beta,i-1}}^{\infty} \frac{4\pi J_\nu(r)}{h\nu} a_\nu d\nu, \qquad (13.5)$$

where $J_\nu(r)$ is the mean radiation intensity at radius r from the central star and a_ν is the photoionization cross section (see Section 2.1). Ionization from excited states can be neglected since neither recombination nor collisional excitation can maintain an appreciable population at excited states. Since H has only one ionization state, the equation for H ionization is

$$\frac{\partial x}{\partial t} + v\frac{\partial x}{\partial r} = (1 - x)\Gamma - n_e x\alpha_A. \qquad (13.6)$$

Time-dependent ionization models were first constructed by Tylenda (1983). As the effective temperature of the star increases, the emission rate of Lyman continuum photons is changing rapidly (see Fig. 13.4). This results in a strong heating of the nebular gas that is not balanced by cooling, resulting in the rise of the electron temperature behind the ionization front. Tylenda (1983) also suggests that the Zanstra method can

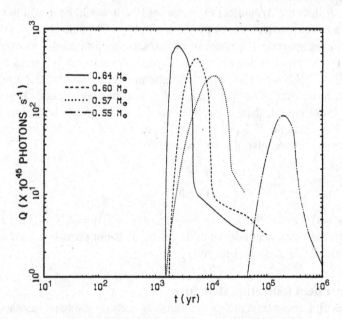

Fig. 13.4. Change of the Lyman continuum photon output with time for four CSPN masses using Schönberner evolutionary tracks (from Kwok, 1985b).

underestimate the central star temperature if the recombination rate falls below the rate of photoionization.

Time-dependent ionization structures for H and He are included in the dynamical models of Schmidt-Voigt and Köppen (1987a), Marten and Schönberner (1991), and Frank and Mellema (1994b). The ionization structures for carbon, nitrogen, and oxygen are assumed to be in equilibrium. The equations become more complicated as more ions and more ionization stages are included, but this may be necessary because many of the metal lines are commonly used for nebular diagnostics and chemical abundance determinations.

13.4 Heating and cooling

Radiative cooling can be included in the dynamics by adding a term in the energy equation [Eq. (12.9)]. Simple cooling functions were incorporated into the models of Volk and Kwok (1985) and Bedogni and d'Ercole (1986). In general, the electron temperature as a function of r and t can be determined from the heating and cooling rate and the ideal-gas-law equation of state. Heating (by photoionization) and cooling (by recombination, forbidden lines, bremsstrahlung, etc.) processes are included in the dynamical models of Schmidt-Voigt and Köppen (1987a), Marten and Schönberner (1991) and Frank and Mellema (1994b). Frank and Mellema estimate that the velocity of the outer shock can be reduced by ~40% because of energy losses by cooling of the hot bubble.

The hot bubble is expected not only to emit continuum radiation in the X-ray and extreme UV, but also infrared coronal lines from highly ionized atoms. Examples of such lines are Nev (14.3 and 24.3 μm), MgIV (4.49 μm), and SiVI (1.96 μm). A theoretical X-ray spectrum of PN is shown in Fig. 13.5.

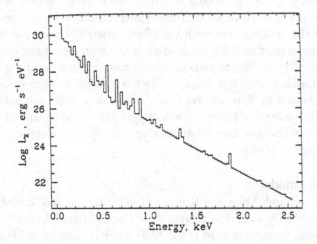

Fig. 13.5. Theoretical X-ray spectrum of a PN as predicted by the ISW model (from Zhekov and Perinotto, 1996).

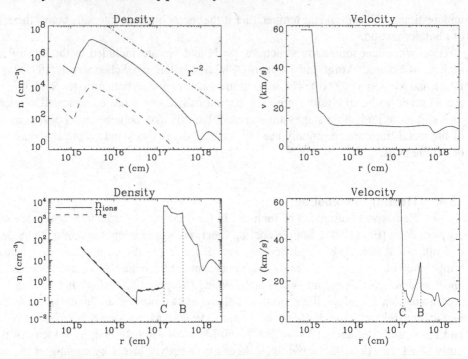

Fig. 13.6. Evolution of the density and velocity structure from the end of the AGB (top panel) to a well-developed PN (4,603 yr after the AGB). The initial density profile reflects the mass loss history used in Blöcker (1995a,b). The region between B and C is the PN shell. Note that the outer shock (B) is expanding much faster than the shell (from Schönberner *et al.*, 1997).

13.5   Expansion velocity of PN

The ISW model also gives us new insights into the interpretation of the expansion of PN. The dynamical age of the nebula is derived from the expansion velocity, which is defined by the position of the shock front. However, this velocity cannot be measured spectroscopically, and the expansion velocities commonly quoted in the literature in fact refer to the matter velocity at radial positions with the largest emission measures. Observationally, the expansion velocities are measured from the line splitting or the FWHM of an emission line (Weinberger, 1989). Simulated profiles calculated from hydrodynamical models by Schönberner *et al.* (1997) show that such velocities systematically underestimate the true expansion velocity by as much as a factor of 2 (see Fig. 13.6). The discrepancy between the evolutionary age and dynamical age discussed in Section 11.5 is therefore resolved.

13.6   The three-wind model

The models of Schmidt-Voigt and Köppen (1987), Marten and Schönberner (1991), Frank (1994) and Mellema (1994) assume three phases of mass loss: (i) a slow wind that has been flowing throughout most of the AGB phase is changed to (ii) a slow wind of high rates (called the superwind) lasting several thousand years; and (iii) the fast wind is initiated several thousand years after the termination of the superwind. The interactions between the three winds result in a more complicated dynamical structure of the PN system. In addition to the forward shock that is driven into the superwind by the thermal pressure of the hot bubble and the contact discontinuity that separates the hot bubble from the superwind, there is also another shock at the superwind-AGB wind interface, which Frank terms the "boundary shock." This boundary shock will be accelerated and can sweep up enough material from the AGB wind to form a new shell. As the central star evolves and its UV output changes, the observational appearance of PN can undergo dramatic changes as the result of the dynamical/ionizational evolution. Frank (1994) divides these developments into four phases: (i) preionization, (ii) H ionization, (iii) He ionization, and (iv) shock breakout. The plots of gas density, velocity, and temperature as a function of radius are given in Fig. 13.7. The emergence of a "double shell" can be seen in Fig. 13.7.

The added complexities in the three-wind variant of the ISW model provide a promising explanation to the multiple shells observed in PN and their diverse kinematics (Chu *et al.*, 1987; Chu, 1989). The existence of multiple shells, which once seemed to demand the hypothesis of multiple ejection (Trimble and Sackman, 1978; Tuchman and Barkat, 1980), can be accounted for by dynamical effects alone. For example, Chu found that some outer shells of PN are expanding faster than the inner shell. Figure 13.6 shows that exists an outer rim with high expansion velocity. The same effect can be seen in Fig. 13.6, which uses the mass-loss history of Blöcker (1995a, 1995b). An outer shell (just inside point B) is expanding faster than the main nebula shell (just beyond point C).

One should not, however, take the name three-wind models too literally. Although the mass-loss mechanism for central stars of PN (hot and compact objects) is clearly different from that of AGB stars (cool and large objects), there is no clear qualitative difference between the ordinary AGB wind and the superwind. Rather, we are witnessing a continuously changing mass-loss rate and ejection velocity as the star ascends the AGB. Because of the numerical complexities of tackling the dynamics of such a system, the "three-wind" model can be considered as a useful approximation.

13.7 Derivation of nebular properties from dynamical models

Since many of the physical properties of PN have been derived assuming a single-shell ionization-bounded nebula, these methods need to be re-evaluated with the ISW model. For example, the ionized masses of PN are traditionally derived from the Hβ flux by using Eq. (4.37). Although this formula will give a reasonable approximation to the ionized mass if the nebula is optically thick to H-ionizing radiation, it underestimates the total ionized mass when the nebula becomes optically thin (Fig. 13.8).

A similar problem exists for the Zanstra method for determining the central-star temperature. Figure 13.9 shows the difference between the actual stellar effective temperature and the simulated H and He Zanstra temperatures. Again, after the break out of the ionization front, the Zanstra temperatures consistently underestimate the true temperature of the star.

It is clear that reliable nebular parameters such as mass and expansion velocity, as well as properties of the central star can only be obtained by employing a dynamical model.

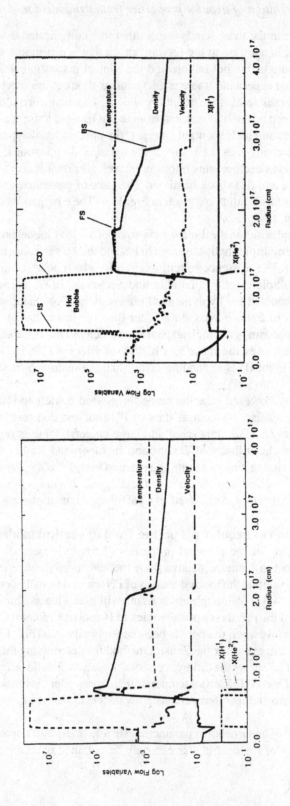

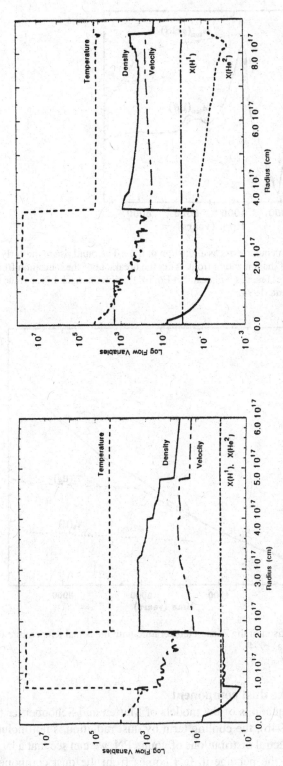

Fig. 13.7. Nebular structure in the three-wind model during the (upper left) preionization ($t = 1,100$ yr);(upper right) H-ionization ($t = 2,000$ yr); (lower left) He-ionization ($t = 3,500$ yr); and (lower right) shock breakout phase ($t = 5,000$ yr). The density (solid line), temperature (dotted line), and velocity (dash line) are shown logithmically as a function of radius. Also plotted are the fractional ionization of H and He (Figures from Frank, 1994).

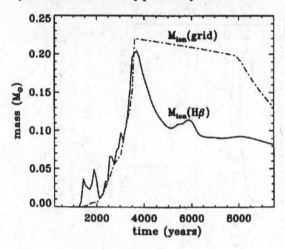

Fig. 13.8. Ionized nebular mass as a function of time. The rapid rise in the early phase is due to the expansion of the ionization front. After the break out of the ionization front, the actual mass (dashed curve) reaches a plateau but Eq. (4.37) gives a much lower value (solid curve; figure from Mellema, 1994).

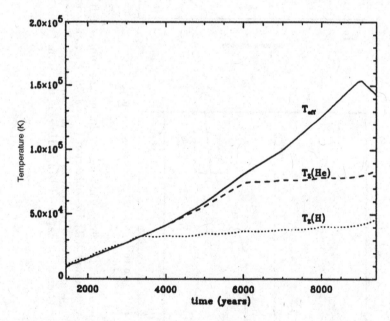

Fig. 13.9. Comparison of the Zanstra temperature with the actual temperature of the central star (from Mellema, 1994).

13.8   Evolution of the dust component

In the energy equations of the models of Marten and Schönberner (1991) and Frank and Mellema (1994b), the cooling term by dust radiation is not included. However, from the energy spectral distributions of young PN, we can see that a large fraction of the energy output of the nebulae in fact comes from the dust component (Zhang and Kwok, 1991). The first attempt to include the dust radiation in the energy balance budget of PN evolution was done by Volk (1992). In this model, the dust component represents the remnant of the wind from the AGB progenitor star, and it is assumed to be dispersing at a constant speed. The dust is heated by the combination of direct starlight and the recombination lines (in particular Lyα) of the ionized component. However, the dust component is assumed to be dynamically decoupled from the gas component and is not part of the dynamical equations. Figure 13.10 shows the spectral evolution of PN from the end of AGB mass loss (upper left), to the dust shell becoming optically thin and the reemergence of the star from the dust shell (lower left), through the onset of photoionization (upper right), to a fully ionized nebula (lower right). The energy is gradually shifted from being radiated in the dust component to the gas component.

13.9   Shaping of planetary nebulae

In the one-dimensional ISW models presented in Section 12.2, both the fast wind and the slow wind are assumed to be isotropic and the resultant PN will have spherical symmetry. In that case, the speed of the swept-up shell can be determined from simple hydrodynamic principles through energy, momentum, and mass conservation at the interface of the fast wind and the AGB wind. The geometrical thickness of the swept-up shell can be obtained if the shell's density law and the density jump between the swept-up shell and the undisturbed AGB material are known. If any one of the fast wind, slow wind, or radiation field is not isotropic, then the nebula will take on a shape reflecting the nature of the asymmetry. Of the three, the slow wind seems to be the most likely candidate to carry significant asymmetry, although no accepted theory exists that explains the origin of the asymmetry (Livio, 1995). Mechanisms that have been proposed for the formation of the density contrast include: (i) protostellar disk left over from the star formation process; (ii) an equatorially compressed outflow; (iii) stellar rotation; (iv) effects of stellar magnetic fields; and (v) the action of a binary companion (Livio, 1994).

The following density profile is adopted by Kahn and West (1984):

$$\rho(\theta) = \rho(0)(1 + \epsilon \sin^n \theta), \tag{13.7}$$

where $\rho(0)$ is the density at the pole ($\theta = 0$), $(1 + \epsilon)$ is the equator-to-pole density ratio, and n is a parameter that determines the gradient of density change from equator to pole. When $\epsilon = 0$, this expression reduces to the spherically symmetrical case. The larger the values of ϵ and n are, the more the AGB wind material is concentrated around the equatorial plane.

From such AGB density profiles, Kahn and West (1985), Icke (1988), and Soker and Livio (1989) demonstrated that the inner boundary of the swept-up shell could range in shape from ellipsoidal to a "figure 8". The interacting winds process will allow the swept-up shell to move faster in the direction of least resistance (along the poles): therefore it further magnifies the original density contrasts between the equator and the poles. Balick (1987) created a two-dimensional morphological scheme which allows nebulae of round, elliptical, and butterfly classes to represent the early, middle, and late stages of the ISW process. Icke (1988) illustrates that for cylindrical symmetry the shock evolves

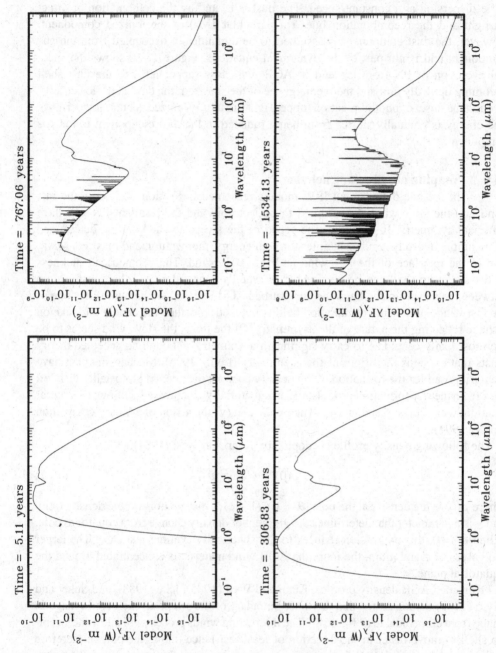

Fig. 13.10. Evolution of the spectrum of PN for a 0.64 $M_\odot$ star. The temperatures of the central star are 5,000, 5,380, 17,800, and 101,700 K at times of 5,307, 767, and 1,534 yr after the AGB (from Volk, 1992).

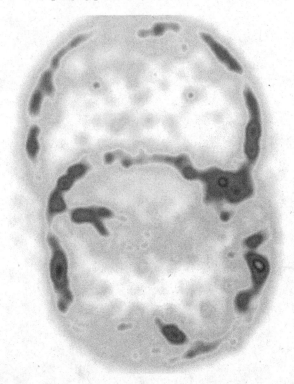

Fig. 13.11. Simulated image of a planetary nebula viewed at an inclination of 60° above the equatorial plane. This image is calculated with the ISW model, using the smoothed particles hydrodynamics method (Li *et al.*, 1999).

toward a fixed shape in a finite time, after which the shock changes only by expanding with a constant scale factor. If this time is very short (e.g., a few hundred years), then the shapes of young PN should find their counterparts among the more evolved larger PN. If cylindrical symmetry is not maintained in real nebulae, conspicuous structural differences may become noticeable between these two groups of objects.

Two-dimensional numerical hydrodynamical calculations for the interaction between an isotropic fast wind and an asymmetric AGB wind have been performed by Frank *et al.* (1993), Frank and Mellema (1994a), Mellema and Frank (1995), and Mellema (1995). Photoionization and radiation mechanisms are included in these models, and the resultant simulated images give a good resemblance to the observed images of PN. The changing UV photon output of the of the central star as it evolves and its effect on the ionization structure and morphology of PN are incorporated in the model of Mellema (1997). An alternative approach to hydrodynamics models is to use the smoothed particle hydrodynamics algorithm, which calculates the density, temperature, and velocity structures from first principles and is particularly well suited to handle asymmetric structures and sudden jumps in the physical parameters. Figure 13.11 shows a simulated image calculated with this method.

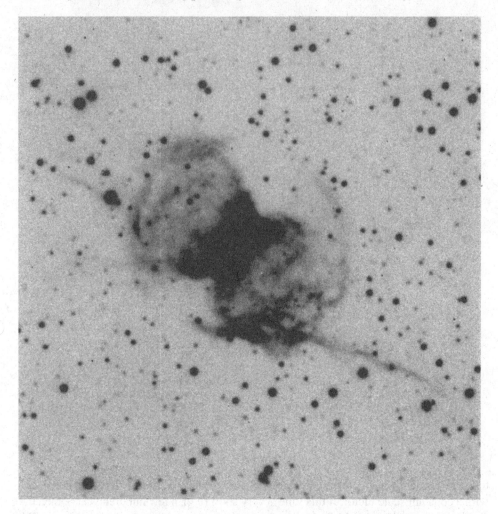

Fig. 13.12. The [NII] image of Sh1-89 (PNG 089.8-00.6; from Hua, 1997).

13.10   Asymmetric ionization structure of PN

If the ISW process is able to amplify the density contrasts between the equatorial and polar directions as the PN evolves, this will also affect the ionization structure of the nebula. The ionization front will propagate faster along the polar directions where the density is lower. This will further dramatize the optical appearance of a PN. Figure 13.12 shows that the PN Sh1-89 consists of a bright core and two faint lobes. It is clear that the core is ionization bounded and the ionizing photons are escaping from the polar directions into the two lobes.

Figures 13.13 show the [SII] and [NII] images of NGC 40. In the [SII] image, one can see that the nebular shell has a sharp outer boundary along the minor axis, giving the impression that the shell is ionization bounded in that direction. In the [NII] image, where fainter emissions are displayed, spherical halos can be seen surrounding the

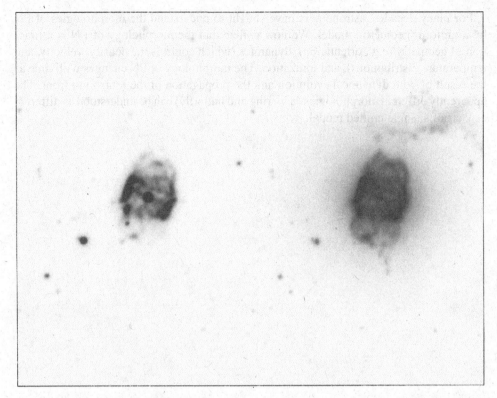

Fig. 13.13. The narrow-band [SII] (left) and [NII] (right) images of NGC 40 (from C. T. Hua).

shell as well as ansae along the major axis. It is clear that the fast wind has pushed much further along the major axis, and the entire nebula (shell and halo) is completely ionized.

13.11   Summary

New hydrodynamical models have led to an overhaul of the interpretation of PN. Halos, double shells, and even multiple shells are natural consequences of the interacting winds process. They do not represent multiple ejections, nor are they related to recurrent He-shell flashes. PN are not simply red giant atmospheres dispersing into the interstellar medium. Photoionization and wind interaction have completely reshaped the circum-stellar environment after the AGB, and the resultant kinematic structure of PN bears no resemblance to the mass ejection history on the AGB.

We have also realized that the interpretation of PN expansion based on the classical uniform-density shell model is simplistic and leads to incorrect expansion velocities and dynamical ages. The traditional methods of estimating nebular masses and central star temperatures by the nebular fluxes are also found to be unreliable. The ISW model also opens the possibility for the first time to understand the diverse morphologies of PN. Even a small density gradient from the equator to pole in the AGB can be amplified to create elliptical, bipolar, butterfly, or double-ring nebulae.

For many decades, astronomers have sought to understand the morphologies of PN by a uniform geometric model. We now realize that the morphology of PN is a function of geometry (e.g., orientation), dynamics (which controls the density, velocity, and temperature distributions), and ionization. The morphology of PN changes with time as the result of both dynamical evolution and the propagation of the ionization front. The apparently different morphologies (e.g., ring and butterfly) can be understood as different aspects of a single unified model.

14

Protoplanetary nebulae – the transition objects

The evolutionary stage between the end of the AGB and PN phases has long been a missing link in our understanding of single-star evolution. As we discussed in Chapter 10, the AGB is terminated by the depletion of the H envelope by mass loss, and this occurs before the onset of carbon detonation. When mass loss reduces the mass of the H envelope (M_e) below a certain value ($M_e \sim 10^{-3}$ $M_\odot$ for a core mass [M_c] of 0.60 $M_\odot$, Schönberner 1983), the star will begin to evolve toward the blue side of the H-R diagram. The effective temperature of the star will increase as the remaining H envelope continues to diminish by H-shell burning. This phase will last until the central star is hot enough ($T_* \sim 30,000$ K) to ionize the circumstellar nebula. The emergence of recombination lines of H, He, and forbidden lines of metals will make the nebula easily observable in the visible, signaling the beginning of the PN phase. A sketch of the evolutionary tracks of proto-planetary nebulae (PPN) in the H-R diagram is shown in Fig. 14.1.

PPN can be defined as the stage of evolution in which their central stars have stopped the large-scale mass loss on the AGB, but have not evolved to be hot enough to emit a sufficient quantity of Lyman continuum photons to ionize the surrounding remnants of the AGB envelope. Observationally, PPN are expected to have the following properties:

(i) PPN should show strong infrared excesses and circumstellar CO emission, suggesting that the remnant of the AGB envelope is still present.

(ii) PPN central stars should have effective temperatures between AGB stars and CSPN. Most of the detected PPN are of spectral class F or G and luminosity class I.

(iii) PPN should show some evidence that the dust envelope is detached from the photosphere and the AGB phase mass loss has ended. Observationally, this is indicated by a double-peak spectral energy distribution.

(iv) Since PPN have neutral envelopes, their nebulae will have no line emission and can only be seen in scattered light. Such reflection nebulae would be difficult to see in the visible against the bright central stars. However, assuming that PPN have similar bipolar morphologies as PN, then for objects that are viewed near edge on where the central star is obscured by dust, the nebula can be detected.

The difficulty in identifying PPN in the past was that many PPN are far infrared sources, and very little was known about the far-infrared sky; and if the PPN were visible, they were difficult to distinguish from ordinary stars. Before the onset of photoionization,

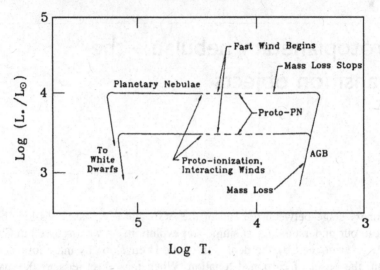

Fig. 14.1. The PPN phase in the H-R diagram. The two tracks correspond to central stars of two different masses.

the nebula was only visible by means of scattered light, which is often very weak. The situation changed as a result of the *IRAS* sky survey in 1983. Many far-infrared sources were detected in the Galaxy, and the hunt for PPN began almost immediately afterward.

14.1   Early PPN candidates

The first PPN candidates, AFGL 618 and AFGL 2688, were discovered as the result of ground-based follow-up observations of objects in the *Air Force Sky Survey (AFGL)*. AFGL 618 is an infrared source of color temperature $\sim$200 K located between two optical lobes separated by $\sim$7 arcsec in the east-west direction (Westbrook *et al.*, 1975). Visible spectropolarimetry showed the two lobes to be reflection nebulae (Schmidt and Cohen, 1981) illuminated by a central star of spectral type B0. Radio continuum observations revealed a small ionized region of $0.4'' \times 0.1''$ embedded in the infrared source (Kwok and Bignell, 1984). The entire optical nebula is surrounded by a molecular envelope of $\sim$20$''$ that is expanding at 20 km s^{-1}, similar to circumstellar envelopes of AGB stars (Lo and Bechis, 1976; Bachiller *et al.*, 1988). From the radial velocity structure, Carsenty and Solf (1982) determined that the bipolar axis is inclined to the plane of the sky at an angle of $\sim$45$°$. The amount of extinction (A_V) in the nebula was estimated to be $\sim$70–110 mag (Lequeux & Jourdain de Muizon, 1990; Latter *et al.*, 1992).

AFGL 2688 (Egg Nebula) was discovered by Ney *et al.* (1975) and is excited by a central star of spectral type F2-5 I (Crampton *et al.*, 1975). Its optical bipolar lobes extend over 10$''$ and have a velocity difference of $\sim$40 km s^{-1} (Cohen and Kuhi, 1980). The fact that the northern lobe is 4–5 times brighter than the southern lobe suggests that the northern lobe is tilted toward us, and the scattered light from the southern lobe is attenuated by the dense dust in the equatorial region. Two pairs of radial "searchlight beams" can be seen emerging from a dark lane in the center. Concentric rings separated by $\sim$0.6$''$ to 1.8$''$ can be seen in the optical image (Sahai *et al.*, 1998a). Assuming an expansion velocity of 20 km s^{-1} and a distance of 1 kpc, the time interval between the rings is ~150–450 yr. Since the interpulse period for AGB stars is >10^4 yr [see Eq. (10.2], thermal pulses cannot be responsible for the production of these arcs. Millimeter-wave interferometric observations show that the CO envelope is centered on the infrared source and extends over 30″ (Heiligman, 1986). ^{13}CO maps made by Yamamura *et al.* (1996) do not show any departure from spherical geometry, suggesting that the AGB mass loss was isotropic. From the density profile of ^{13}CO, Yamamura *et al.* (1996) concluded that the mass-loss rate of the star increased from 10^{-4} to 3 × 10^{-4} $M_\odot$ yr^{-1} over a 3,000-yr period. The total amount of molecular mass is estimated to be ~0.9 $M_\odot$, similar to that observed in NGC 7027 (see Section 5.4).

Molecular hydrogen was first observed by Beckwith *et al.* (1978), and subsequently four clumps of H$_2$ emissions were found (Gatley *et al.*, 1988; Latter *et al.*, 1993). The four clumps form a remarkable crosslike pattern, which is best displayed in the HST NICMOS image (Sahai *et al.*, 1998b). The clumps in the N and S directions of AFGL 2688 can be seen to coincide with the brightest regions of the optical emission, whereas the E and W clumps are aligned along the obscuring equatorial disk. The strengths of the H$_2$ lines observed are consistent with shock excitation, and the H$_2$ clumps probably trace the inner surfaces of the AGB wind that are impinged by the fast wind (Cox *et al.*, 1997).

The visual brightness of both AFGL 618 and AFGL 2688 has increased since before 1920. From archival photographs, Gottlieb and Liller (1976) found that the *B* mag has been decreasing at a rate of ~0.06 and ~0.05 mag yr^{-1} for AFGL 618 and AFGL 2688, respectively.

The most plausible explanation of the nature of AFGL 618 and AFGL 2688 is that they are post-AGB objects evolving toward the PN stage. The infrared and molecular emissions originate from the remnants of the circumstellar envelopes of their AGB progenitors. Shortly after their departure from the AGB, possibly as the result of an interacting winds process (see Section 13.9), their envelopes become increasingly asymmetric. While the central stars remain heavily obscured by circumstellar dust in the equatorial plane, the lower densities and smaller optical thickness at the poles allow the visible light to escape, and the scattered light forms a bipolar nebula. Although the distances to these objects are uncertain, their minimum luminosities derived from the infrared fluxes [1.2 × 10^4 and 1.4 × 10^3 (D/kpc)^2 $L_\odot$ for AFGL 618 and AFGL 2688 respectively; Hrivnak and Kwok, 1991], are consistent with their being post-AGB objects. The increase in visual brightness is probably due to a decreased extinction as the dust envelope disperses. In AFGL 618, the central star is probably more massive (therefore evolving faster) and is now just hot enough to begin to ionize the circumstellar environment. A steady increase in the f-f continuum flux level of AFGL 618 in the 21-cm to 800-μm range suggests that the ionized region began to form approximately 20–50 yr ago (Kwok and Feldman, 1981; Knapp *et al.*, 1993).

14.2 The Search for PPN

In Chapter 10 we discussed that as stars ascend the AGB, their mass-loss rates increase and the stars become redder and redder as more and more of their light is absorbed by the dust component. A plot of AGB stars together with PN on an *IRAS* color-color diagram (Fig. 14.2) shows that PN have even redder colors. We can see that the colors of O-rich AGB stars and young PN are clearly separated. In fact, a general trend of decreasing color temperature can be observed starting from silicate emission objects, to silicate absorption objects, to PN with high surface brightness, to PN with low surface brightness. This apparent evolutionary sequence in the *IRAS* color-color diagram is the result of the monotonic decrease of the dust temperature from the AGB to the PN phase. On the AGB, where the mass-loss rate is increasing, the change to lower color temperatures is the result of increasing optical depth in the dust circumstellar envelope. After mass loss has terminated, the decrease in color temperature is caused by geometric dilution as a consequence of the remnant envelope is expanding away from the star (Kwok, 1990).

Since PPN are expected to have *IRAS* colors between those of late AGB stars and young PN, candidates for PPN can be identified by searching the *IRAS* (PSC) for objects with the appropriate colors (Volk and Kwok, 1989; van der Veen *et al.*, 1989; Slijkhuis, 1992). Generally speaking, we expect the color temperature of PPN to be in the range 150–250 K.

Approximately 1000 PN were detected in the *IRAS* survey. According to the evolutionary models of Schönberner (1983), the fraction of time spent in the PPN phase in relation to the entire PN lifetime is $\sim$10%. If this is the case, we expect $\sim$100 PPN in the *IRAS PSC*. These candidates can be identified in two ways: either by looking for stars in existing optical catalogs with appropriate *IRAS* colors, or by searching for the optical counterparts of low-temperature *IRAS* sources. The PPN candidates discovered to date have been found through both of these two strategies.

14.2.1   PPN candidates associated with known optical stars

The first systematic effort in searching for stars in existing optical catalogs with cool dust shells was carried out by Bidelman (1985, 1986). By identifying stars with known optical spectral types with *IRAS* fluxes peaking at 25 μm, Bidelman found several stars of intermediate spectral types (F-K) with large infrared excesses that might be stars in the post-AGB phase of evolution. Among these objects are a number of high-latitude F supergiants which show high space velocity, low metal abundance (Bond and Luck, 1987), circumstellar dust (Parthasarathy and Pottasch, 1986), and molecular envelopes (Likkel *et al.*, 1987). The supergiant spectral classification reflects the low surface gravity of these stars, but not their intrinsic luminosity. The prototype of this class is 89 Her, which is a MK spectral standard of type F2 Ia. 89 Her is located at a galactic latitude of 23°, which puts it at an unreasonably large distance above the plane if it has the luminosity of a Population I supergiant. Additional support for its post-AGB status comes from its infrared excess (Gillett *et al.*, 1970), which could have originated from the remnant of the circumstellar envelope of its AGB progenitor.

Examples of high-latitude luminous stars that have been suggested as post-AGB objects, based on their infrared excesses, include HD 161796 (IRAS 17436+5003, F3 Ib, $b = 31°$) and HD 101584 (IRAS 11385–5517, F0 Iep, $b = 6°.0$) by Parthasarathy and Pottasch (1986), HR 4049 (IRAS 10158–2844, B9.5 Ib-II, $b = +22°.9$) by Lamers *et al.* (1986), and HD 213985 (IRAS 22327–1731, A2 Ib, $b = -57°$) by Waelkens *et al.* (1987).

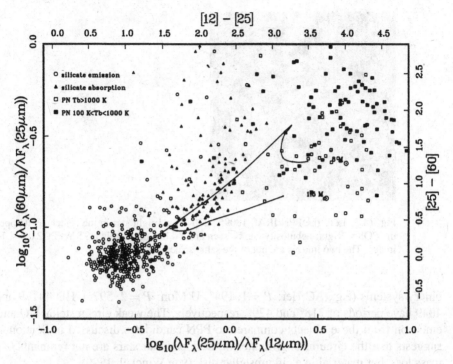

Fig. 14.2. The distribution of AGB stars and planetary nebulae in the color-color diagram. The open circles are AGB stars showing the silicate feature in emission, and the triangles are AGB stars showing the silicate feature in absorption. The open squares are young PN with a 5 GHz brightness temperature (T_b) > 1,000 K, and the filled squares are more evolved PN with 1000 K > T_b > 100 K. Also plotted are two PPN evolutionary tracks (for MS masses of 1.5 and 8 $M_\odot$, respectively).

Two other classes of variable stars have been suggested as post-AGB stars. RV Tauri stars are characterized by light curves showing alternate deep and shallow minima, with periods of 50–150 days, and have spectral type F, G, or K. The circumstellar envelopes of RV Tauri stars can also be seen in CO (Bujarrabal *et al.*, 1988; Alcolea and Bujarrabel, 1991). The class of UU Her stars is described by Sasselov (1984) as small-amplitude variable stars with periods of 40–100 days that are located at high galactic latitudes. Infrared excesses similar to 89 Her have been found in RV Tauri (Gehrz, 1972), but not in UU Her (Gehrz and Woolf, 1970). A comprehensive study of 25 high galactic latitude supergiants with infrared excesses was made by Trams (1991).

There are a number of other peculiar infrared sources that have been suggested as stars in the post-AGB phase of evolution. The most well-known examples are the HD 44179 (the Red Rectangle; Cohen *et al.*, 1975) and the Frosty Leo nebula (IRAS 09371+1212; Forveille *et al.*, 1987). HD 44179 is a star of uncertain spectral type between B9 and F that shows strong UIR features (Russell *et al.*, 1978). The Frosty Leo nebula has a large far-infrared excess that is due to the 45-μm ice band (Omont *et al.*, 1989). The ice band also shows up as a prominent absorption feature at 3.1 μm (Geballe *et al.*, 1988).

The discovery that many of these peculiar stars are binaries brings doubts to whether they are genuine post-AGB stars. The best known RV Tauri stars are all members of wide

Fig. 14.3. Left: the PPN IRAS 18095+2704 (the bright star in the center) as it appears on the POSS. Right: nebulosity can be seen around the star in this HST WFPC2 0.4 s, *V*-band image. The two images are not of the same scale.

binary systems (e.g., AC Her: $P = 1,194^d$, U Mon: $P = 2,597^d$). HD 44179 and HD 4049 have periods of 318^d and 429^d, respectively. The weak circumstellar CO and dust emission from these objects compared to PPN candidates discussed in Section 14.2.2 suggests that the circumstellar matter in these peculiar stars are not remnants of AGB mass loss, but material in a circumstellar disk (van Winckel, 1999).

14.2.2   Cool IRAS sources as PPN candidates

Another method of finding PPN is to seek optical counterparts of *IRAS* sources satisfying certain color criteria. Candidates are selected from a certain region of the *IRAS* color-color diagram (Fig. 14.2) and are identified with a ground-based telescope or by positional coincidence on sky survey plates. This search strategy for PPN depends critically on the correct identification of the optical counterpart of the *IRAS* source. Many *IRAS* sources are evolved AGB stars that suffer from large circumstellar extinction and thus will not be visually bright (Kwok *et al.*, 1987). The association of stars in optical survey plates with the *IRAS* source by pure positional coincidence can lead to the wrong identification of the *IRAS* source, particularly in the galactic plane where the field is crowded. Even when a counterpart is identified by searching in the *K* band, the possibility for confusion still exists because most red stars are strong near-infrared emitters. The only sure way to obtain the correct identification is by searching around the *IRAS* position at 10 or 20 μm and comparing the observed fluxes with the *IRAS* 12- and 25-μm PSC fluxes. Although the *IRAS* positions are generally good, there are cases in which the actual source can be up to 1 arcmin from the PSC position.

One of the first PPN discovered by this method is IRAS 18095+2704 (Hrivnak *et al.*, 1988). It is identified with a 10 mag star of spectral type F3 Ib at the galactic latitude of $b = +20°.2$ (Fig. 14.3). The most outstanding characteristics of the SED of this object is that it shows two distinct components, each contributing significantly to the total observed flux. The cooler component corresponds to the remnant of the AGB dust envelope, and the warmer component corresponds to the reddened photosphere of the central star. A number of other PPN candidates are found to have similar behavior. Two examples of

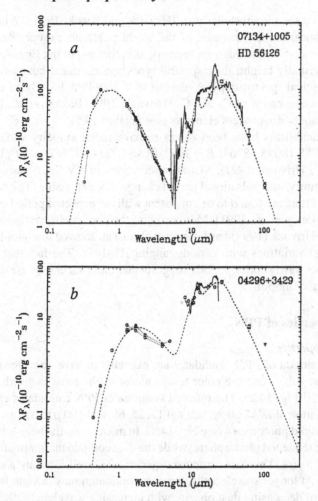

Fig. 14.4. The SEDs of *a*, IRAS 07134+1005; and *b*, IRAS 04296+3429.

the SED of PPN are shown in Fig. 14.4. The clear separation of the two flux components suggests that the dust envelope is detached from the photosphere, which is consistent with the theoretical expectation that the AGB mass loss terminated some time ago (Fig. 13.10). The dynamical age (after the cessation of mass loss) of these objects can be estimated by radiative transfer models, and the typical time scales of the PPN range from several hundred years to more than a thousand years (Hrivnak *et al.*, 1989).

14.3 Optical properties of PPN

Unlike PN, the optical light from PPN is not due to nebular emission but is primarily due to the stellar photospheric continuum plus contributions from scattered light from the nebula. Because of the presence of circumstellar dust, the stellar light is often strongly reddened. The PPN discovered to date vary greatly in apparent magnitude, ranging from ~7 mag to >22 mag. This can be a result of different orientations in the sky. PPN viewed pole on will appear bright in the visible, whereas those viewed edge on will suffer heavy extinction by the circumstellar dust (Hrivnak and Kwok, 1991). Whereas pole-on systems will appear starlike because of the bright photospheric contribution, reflection nebulae can be seen around edge-on systems, as in the case of the Egg Nebula.

For objects that are visually bright, their spectral types and chemical compositions can be determined by optical spectroscopy. A number of C-rich PPN have been found by the presence of molecular bands of C_2 and C_3 (Hrivnak, 1995; Bakker *et al.*, 1997) as well as an overabundance of s-process elements (see Section 19.5).

A number of PPN candidates have been found to have radial velocity variations caused by pulsations (IRAS 18095+2704: $P = 109^d$, IRAS 2222+4327: $P = 89^d$, IRAS 22272+5435: $P = 127^d$; Hrivnak, 1997). Visual observations of RV Tauri stars have been made for over a century, and pulsational period changes are detectable. The period changes of two RV Tauri stars are found to be consistent with the expected period evolution of post-AGB stars (Percy *et al.*, 1991). Monitoring of the photometric variations of ~40 PPN candidates by Hrivnak (1997) has found that almost all showed low-amplitude (Δm_V ~0.15–0.35 mag) variations with periods ranging 25–146^d. The fact that PPN with earlier spectral types have shorter periods suggests that these variations are due to pulsation rather than binary motion.

14.4   Infrared properties of PPN

14.4.1   *Continuum properties*

The infrared continua of PPN candidates are expected to have color temperatures of 150–300 K, that is between the color temperatures of the most evolved AGB star and the youngest PN (Fig. 14.2). The infrared continua of PPN candidates can be determined by a combination of *IRAS* photometry at 12, 25, 60, and 100 μm and ground-based near- and mid-infrared photometry (see Fig. 14.4). In most cases, the near-infrared emission arises from the reddened photosphere, while the dust component dominates the continuum longward of 5 μm. The dust component generally is well fitted by a single temperature blackbody. At longer wavelengths, the dust continuum may deviate from a blackbody as the result of decreasing dust opacity with increasing wavelength. Submillimeter continuum observations can help better define the shape of the dust continuum and the mass loss history (van der Veen *et al.*, 1994).

It has been suggested that the near infrared excesses seen in HR 4049 represents mass loss in the post-AGB stage (Trams *et al.*, 1989). Although some PN (e.g., NGC 7027; Russell *et al.*, 1977; IC 418; Zhang and Kwok, 1992) are known to have a hot dust component in addition to a cool dust component from the AGB mass loss, the hot dust seen in HR 4049 is likely not the result of post-AGB mass loss, it probably originates from a stationary circumstellar disk (see Section 14.2.1).

14.4.2   *The silicate feature in O-rich PPN*

Since the 10- and 18-μm features of silicates are commonly observed in O-rich AGB stars, it is expected that these features will also be observable in O-rich PPN. It is expected that the 10-μm feature in PPN will be much less prominent than that in AGB stars because of the decline in dust temperature and the shift of the spectral peak to longer wavelengths. This effect is quantitatively confirmed by the detached-shell radiative transfer model of Volk and Kwok (1989), who found that the 10-μm feature not only weakened but also broadened. A search of the *IRAS LRS* has yielded a number of sources

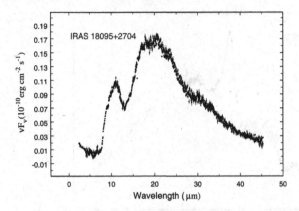

Fig. 14.5. The ISO SWS spectrum of the O-rich PPN IRAS 18095+2704 showing the 9.7 and 18 μm silicate features.

with the predicted shape at 10 μm (Volk and Kwok, 1989). A number of these objects have now been confirmed to be PPN (e.g., IRAS 18095+2704, IRAS 10215−5916, and IRAS 20004+2955). An example of the infrared spectrum of O-rich PPN is shown in Fig. 14.5. The existence of O-rich PPN confirms the evolutionary connection between OH/IR stars and PN.

14.4.3 The UIR features

In addition to the well-known 3.3-μm UIR emission feature that is commonly observed in PN and HII regions (see Section 6.3), there exist emission features in the 3.4 to 3.5-μm region that, although present in PN, are found to be strongest in PPN. These features have been detected in the PPN candidates IRAS 05341+0852 (Geballe and van der Veen, 1990), IRAS 04296+3429, IRAS 22272+5435 and AFGL 2688 (Geballe *et al.*, 1992), as well as the C-rich young PN IRAS 21282+5050 (Geballe and van der Veen, 1990). It has been suggested that the normal 3.4- to 3.5-μm emission peaks seen in PN are due to hot bands of the fundamental 3.3-μm CH stretch (Barker *et al.*, 1987). However, the unusually strong 3.4- to 3.5-μm emission seen in PPN cannot be attributed to hot bands (Geballe *et al.*, 1992, 1994). The lack of UV flux from the central stars of PPN makes excitation of the emitting molecules to high vibrational levels unlikely. It seems more probable that the emissions longward of the 3.3 μm feature are due to vibrations of aliphatic sidegroups (e.g. $-CH_2$, $-CH_3$) attached to PAHs (Jourdain de Muizon *et al.*, 1989).

KAO observations in the 5- to 8-μm region showed that the 6.9- and 8-μm UIR features observed in HII regions and PN are also present in PPN (Buss *et al.*, 1990). Whereas the 3.3- and 6.2-μm features observed in HII regions and PN are thought to be fluorescently excited by UV photons, this cannot be the case in PPN because of their late spectral types. If these bands in PPN are excited by visible photons, then the molecules responsible must be larger (>100 C atoms) than interstellar PAH molecules (Buss *et al.*, 1990). Based on the strength correlation between the 3.4- to 3.5-μm features and the 6.9-μm feature in IRAS 22272+5435, Geballe *et al.* (1992) suggest that the former is due to the stretching mode of CH_2 and CH_3 groups. Because of the different temperatures of the exciting stars in PN and PPN, the strengths of the 3.4- to 3.5-μm relative to the 3.3-μm features are due to the different amounts of UV and visible photons available.

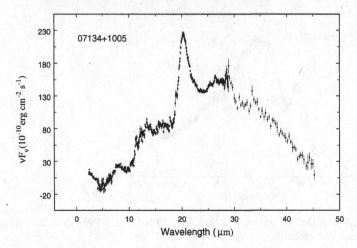

Fig. 14.6. The ISO SWS spectrum of the PPN IRAS 07134+1005 showing the presence of the 21-μm feature.

14.4.4   The 21-μm feature

The dominant circumstellar dust grains in C-rich AGB stars are SiC and, in extreme carbon stars, amorphous carbon (see Section 10.4.1). The strength of the 11.3-μm SiC feature is weaker than the 10-μm silicate feature, and amorphous carbon is featureless in the mid-infrared. These facts have led us to assume that C-rich PPN will not have strong identifying features in the 10- to 20-μm range. The discovery of a strong emission feature at 21 μm in four PPN therefore came as a surprise (Kwok *et al.*, 1989). The *IRAS* LRS of these four sources show a prominent feature at 21 μm with an almost flat (in λF_λ) continuum between 12 and 18 μm (Fig. 14.6). The 21-μm feature has been confirmed by both *KAO* (Omont *et al.*, 1995), ground-based (Kwok *et al.*, 1995; Justtanont *et al.*, 1996), and *ISO* observations Volk *et al.*, 1999).

The detection of a 30-μm emission feature in four 21-μm sources (IRAS 07134+1005, IRAS 20000+3239, IRAS 22272+5435, and IRAS 23304+6147; Omont *et al.*, 1995) suggests that these two features share some common origin. However, the 30-μm feature has been seen in AGB stars (e.g., IRC+10216) and PN (e.g., IC 418) as well as in PPN (Cox, 1993), whereas the 21-μm feature thus far appears to be restricted to PPN.

The detection of strong 3.4- to 3.5-μm features in several of the 21-μm sources is particularly interesting. While the 3.3-, 7.7-, and 11.3-μm features are commonly seen in PN, the only PN that show somewhat comparable 3.4- to 3.5-μm features are BD+30°3639, NGC 7027 (Geballe *et al.*, 1985; Nagata *et al.*, 1988), and IRAS 21282+5050 (de Muizon *et al.*, 1986), all of which have large infrared excesses and are regarded to be very young. However, the profiles of the 3.4- to 3.5-μm features and the wavelengths of peak intensity are not exactly the same in these 21-μm sources as in the PN. Furthermore, the ratios of the strengths of the 3.4–3.5-μm to 3.3-μm features are much larger in these 21-μm sources (Geballe *et al.*, 1992). Since the 3.3-μm feature is caused by aromatic C-H stretch whereas the 3.4-μm feature is caused by aliphatic C-H stretch, this suggests an increase of aromatic bonds (sp^2) relative to aliphatic bonds (sp^3) as the star evolves from PPN to PN.

Table 14.1. List of 21 μm sources and their common properties

Name	21 μm	30 μm	Sp. Ty.	Optical	UIR features	Mol. lines
02229+6208	medium	...	G8-K0 0-Ia	C_2, C_3,	...	CO
04296+3429	strong	...	G0Ia	C_2, C_3, CN	3.3, 3.4-3.5, 7.7,11.3	CO, HCN
05113+1347	medium	...	G8Ia	C_2, C_3, CN	3.3, 11.3	CO
07134+1005	v. strong	medium	F5 I	C_2, C_3, CN	3.3, 6.9	CO, HCN
16594−4656	strong	...	F3I	...	...	CO
19500−1709	v. weak	...	F3I	...	...	CO, HCN
20000+3239	weak	v. strong	G8Ia	C_2, CN	7.7, 11.3	CO, HCN
AFGL 2688	weak	medium	F5 Iae	C_2, C_3, CN	3.3, 3.4-3.5	CO, HCN
22223+4327	medium	...	G0Ia	C_2, C_3, CN	...	CO, HCN
22272+5435	strong	v. strong	G5Ia	C_2, C_3, CN	3.3,3.4-3.5, 6.9, 7.7, 11.3	CO, HCN, CS
22574+6609	medium	...	...	...	7.7, 11.3	CO
23304+6147	v. strong	v. strong	G2Ia	C_2, C_3, CN	7.7, 11.3	CO, HCN

Table 14.4.4 lists the 21-μm sources that have been detected as of 1998. These sources have remarkably uniform properties. They are all post-AGB stars with strong infrared excesses and have optical spectral types mostly classified as F and G supergiants. The presence of C_2 and C_3 suggests that these are extremely carbon-rich sources. Many candidates have been proposed as possible carriers of the 21-μm feature, including:

- amides: urea or thiourea (Sourisseau *et al.*, 1992)
- hydrogenated amorphous carbon (HAC; Buss *et al.*, 1990)
- hydrogenated fullerenes ($C_{60}H_m$, m = 1, 2, ...60, or fulleranes; Webster, 1995)
- nanodiamonds (Hill *et al.*, 1998)
- Maghemite (Fe_2O_3) or magnetite (Fe_3O_4; Cox, 1991)
- solid SiS_2 (Nuth *et al.*, 1985; Goebel 1993; Begemann *et al.*, 1996)

In view of the fact that the 21-μm feature is seen exclusively in C-rich objects, it seems unlikely that it is due to inorganic materials such as SiS_2 or Fe_2O_3. The correlation of the 21-μm feature with the UIR features suggests that its carrier is a large carbon-based molecule. Considering that fullerenes (C_{60}) are the three-dimensional analogues of the two-dimensional PAH molecules, their existence in C-rich objects is not unexpected. Emission spectra of gas phase C_{60} show strong features at 7.1, 8.6, 17.5, and 19 μm, whereas the lowest vibrational mode of $C_{60}H_{60}$ is calculated to be at 23 μm (Webster, 1995). For the other unsaturated hydrides ($C_{60}H_m$, m = 1 to 59), the perfect symmetry of the molecule is broken and the mode is no longer a narrow feature but bands peaking at wavelengths from 19 to 23 μm as m changes from 0 to 60. Webster (1995) suggests that the observed broad 21-μm feature is the result of a mixed population of fulleranes.

As a star evolves from the AGB to PN, its circumstellar environment (density, temperature, and radiation background) undergoes drastic changes. The dominant infrared emission features will likely change as the conditions for molecular formation, destruction, and excitation change. Acetylene (C_2H_2), widely considered to be the first building block of PAH molecules, has recently been detected in the atmospheres of carbon stars by *ISO*. Thus it is possible that PAH molecules are synthesized in AGB stars, but that the molecules are not excited because of the low temperatures of the central stars. HAC grains, made up of aromatic rings of various sizes bonded peripherally to polymeric or hydrocarbon species (Duley and Williams, 1986), can also be made during the late AGB phase. Extreme carbon stars on the AGB have featureless dust continua (Volk *et al.*, 1992) and these could be due to HAC grains. During the transition from the AGB to PN, these grains may be gradually shattered into smaller PAH molecules, possibly by high-velocity outflows now observed in PPN. Then as the UV radiation background increases as the central star evolves to higher temperatures, the strengths of the PAH features increase. If 21-μm features are due to HAC grains, then the destruction of HAC grains will also imply the disappearance of the 21-μm feature beyond the PPN phase.

14.4.5   *Atomic and molecular lines*

The near-infrared spectrum of PPN candidates, in particular those of F spectral type, are dominated by hydrogen recombination lines in absorption. For objects of later spectral type (mid-G or later), vibrational bands of CO are generally seen in absorption (Fig. 14.7), although in several cases these bands are in emission. The most interesting case is IRAS 22272+5435 for which the CO spectrum has been seen to change from emission to absorption over a 3-month interval (Hrivnak *et al.*, 1994).

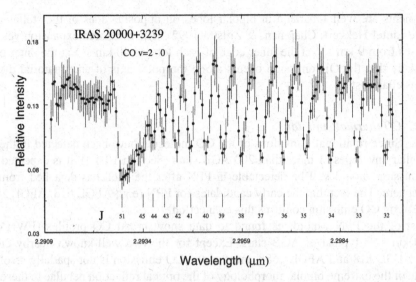

Fig. 14.7. The CO $v = 0 - 2$ vibrational-rotational transitions in the PPN IRAS 20000+3239. The rotational quantum numbers are labeled.

14.4.6 Infrared imaging

Although visible imaging traces the scattered light, mid-infrared imaging has the advantage of actually mapping the dust density distribution. The O-rich PPN HD 161796 and IRAS 19114+0002 were imaged by Skinner *et al.* (1994) and Hawkins *et al.* (1995), respectively. The dust shells in both objects are found to have a ringlike structure, consistent with the picture that they are bipolar nebulae being viewed along the polar directions. The technique of mid-infrared imaging is therefore well suited for the observation of pole-on objects for which optical imaging is difficult because of their bright central stars.

14.5 Circumstellar molecular emissions

14.5.1 OH maser emission

Oxygen-rich AGB stars often exhibit maser emissions in OH, H_2O, and SiO (see Section 10.4.2). However, because of the changing envelope conditions, a long velocity-coherent path length necessary for a saturated maser is difficult to maintain and not all of these maser lines are expected to persist through the PPN stage. Theoretical calculations by Sun and Kwok (1987) suggest that only stars with high mass-loss rates will have a detectable 1612-MHz emission beyond the AGB. An evolutionary scenario of how the strengths of these maser lines will vary after the star has left the AGB is outlined by Lewis (1989). For example, the PPN IRAS 18095+2704 shows OH main (1665/1667 MHz) lines that are stronger than its 1612-MHz line, the opposite of what is usually seen in AGB stars (Lewis *et al.*, 1990). An OH and H_2O survey of cool IRAS sources by Likkel (1989) has detected several PPN candidates (IRAS 17436+5003, IRAS 19114+0002, IRAS 19477+2401, IRAS 23321+6545), all in the main lines.

Of particular interest is the OH 1667 MHz main-line maser emission from the post-AGB star candidate HD 101584 (IRAS 11385−5517). The blue-and red-shifted components are well separated in bipolar lobes on opposite sides of the stellar position (te Lintel Hekkert, Chapman, & Zijlstra 1992). The measured expansion velocity increases from 9 km s^{-1} at the inner edge of each lobe to 40 km s^{-1} at the outer edge. It is likely that the OH emission occurs along the polar axis of an equatorial disk of circumstellar dust.

14.5.2   CO thermal emission

Since rotational transitions of the CO molecule have been detected in the circumstellar envelopes of more than 200 AGB stars (Section 10.4.3), it is expected that CO emission should still be detectable in PPN after the shell has detached from the photosphere. The fact that the early candidates for PPN (e.g., AFGL 618, AFGL 2688) show strong CO emission confirms this connection.

Most of the PPN candidates found to date show broad CO profiles (FWHM of 20–30 km s^{-1}) typical of AGB stars. Except for the two well-known nearby candidates, AFGL 618 and AFGL 2688, most of the CO emission is not spatially resolved. In spite of the extreme bipolar morphology of the optical reflection nebulae in these two objects, the CO envelope is remarkably symmetric (Deguchi, 1995), suggesting that a morphological change occurred after the AGB. The CO envelope of IRAS 19114+0002 was found to be partially resolved by Bujarrabal *et al.* (1992). The envelope shows an elongated structure of approximately $18'' \times 14''$. Future millimeter-wave interferometry observations should produce much improved morphologies of the PPN molecular envelopes and, therefore, allow us to more accurately estimate the fraction of PPN that have nonspherically symmetric envelopes.

One of the major uncertainties in post-AGB evolution is the onset of the fast wind. A significant rate of mass loss during the post-AGB (and pre-PN) evolution will greatly shorten the transition time from AGB to PN (see Section 11.2). Although the PN central-star winds can be detected by P Cygni profiles of the UV-resonance lines, they are difficult to detect before ionization. The only evidence for the existence of a central-star wind during the PPN phase is through the detection of extended wings of the CO line.

The development of SIS receivers has allowed the detection of weak, extended line wings of CO in the PPN AFGL 618 (Gammie *et al.*, 1989; Cernoicharo *et al.*, 1989), AFGL 2688 (Young *et al.*, 1992; Jaminet *et al.*, 1992), and IRAS 19500–1709 (Bujarrabal *et al.*, 1992). Molecular winds, with velocities of approximately ten times the AGB wind velocities, have been found (Fig. 14.8). Such velocities are in the range of expected escape velocities for the spectral types (B and F) of AFGL 618 and AFGL 2688.

14.6   The beginning of photoionization

It would be interesting to identify the exact transition from PPN to PN by detecting the presence of ionized gas in PPN. The ionized component can be detected by infrared recombination lines or by radio continuum emission. IRAS 17516–2525 does not show f-f radiation, but the detection of Brα, Brγ, and Pfγ suggests that this object could be a PPN (van der Veen *et al.*, 1989). The detection of a very small ionized region in the PPN AFGL 618 (Kwok and Bignell, 1984) suggests that this PPN is just about to enter the PN stage.

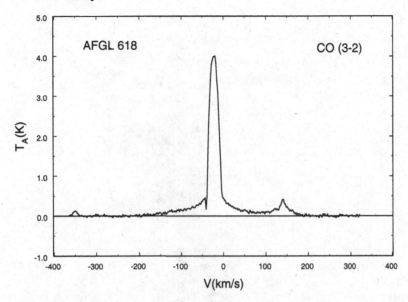

Fig. 14.8. The CO $J = 3 \rightarrow 2$ spectrum of AFGL 618. Extended wings can be seen to ± 200 km s^{-1}.

14.7   Morphologies of PPN

Although the dynamical models have been successful in simulating the diverse morphologies observed in PN (see Section 13.9), the time scale over which certain morphologies develop is uncertain. Does the shaping occur gradually over the lifetime of a PN, or the are morphologies fixed early in the PN's evolution? This question can be answered by the observations of PPN. If PPN already show a similar morphological distribution as evolved PN, then the shaping mechanism must be at work at a very early stage.

An example of a bipolar PPN is shown in Fig. 14.9. We can see a series of concentric rings superimposed on the two reflection nebular lobes, similar to those observed in AFGL 2688 (Kwok *et al.*, 1998). The nearly circular shapes of the rings suggest that mass loss is spherically symmetric during the AGB and the bipolar morphology only develops after the end of the AGB. Such circular rings have also been seen in the bipolar PN such as NGC 7027, Hb 5, and NGC 6543. The emergence of the bipolar structure only a few hundred years after the AGB suggests that the shaping occurs early. This suggests that interacting winds process relying on a spherically-symmetric fast wind may not be adequate to shape the nebula. The tranformation from spherical to bipolar shapes could be due to the onset of a fast collimated outflow initiated at the PPN phase. The direct detection of such outflows would be an important step in the identification of the shaping mechanism of PN.

14.8   Summary

The discovery of PPN has allowed us to bridge the evolutionary gap between the AGB and PN. The dispersal of the molecular and dust envelopes ejected during the AGB phase can now be traced through the PPN phase to the PN stage. The central stars can be

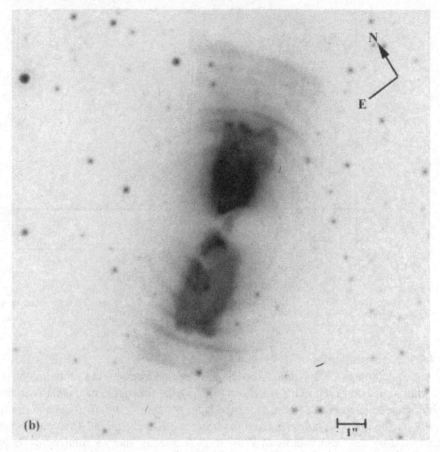

Fig. 14.9. HST WFPC2 image of IRAS 17150−3224.

observed and their chemical abundances determined. The beginning of photoionization can be identified.

PPN also hold the key to many of the questions that we have for the evolution of PN. When does post-AGB mass loss begin? What is the magnitude of this wind and how does it affect the transition time? Is this wind spherically symmetric and is it responsible for the shaping of PN? How early does the morphology of PN develop? When are the hydrocarbon molecules responsible for the UIR features synthesized and excited?

PPN are also interesting in their own right. All intermediate-mass stars will pass through the post-AGB stage, most will evolve into PN, and some (the lower-mass ones) may not. Possibly because of their unique temperature, density, and chemical environments, they are the only objects known to show the 21-μm feature. Other than supergiants that evolved from massive stars, they are the most optically luminous stellar objects. Although galactic PPN were only discovered recently, their visible brightness suggests that extragalactic PPN can be detected and could even serve as standard candles for the determination of cosmological distance scales (Bond, 1997).

15

Evolution to the white dwarf stage

In Chapter 11 we discussed that the central stars of PN originate from the electron-degenerate carbon-oxygen core of AGB stars. After the thin H envelope has been depleted by nuclear burning and mass loss, CSPN will change from deriving their energies from the CNO process to gravitational contraction. This is accompanied by a drop in luminosity by at least an order of magnitude. He burning is unimportant (except during a thermal pulse) and also dies away eventually. The star now enters the "cooling track" with a gradual decline in both luminosity and temperature. These blue and faint stars are referred to as white dwarfs (WDs).

WDs were discovered as faint stars that are unusually dense that perturb the orbits of their normal companions through gravitational interaction. Most of the early WD discoveries are nearby single stars, which are found in surveys of stars with large proper motions (Luyten, 1979). More recent discoveries have resulted from surveys of faint blue stars, such as the quasar survey of Palomar-Green (Green et al., 1986). The PG survey has led to the discovery of hot, H-deficient stars known as PG 1159 stars (see Section 7.3). These are the hottest WDs and therefore can be considered as possible transition objects between the PN and WD stages.

For very low-mass stars, it is possible that mass loss or binary mass transfer on the RGB removes sufficient mass from the envelope to stop nuclear evolution before the He flash. The remnants of such stars will be WDs with He cores (see Chapter 10). In a few stars with masses high enough to be close to the degenerate limit for the C-O core, C ignition may be able to remove the core degeneracy. The subsequent C burning will create a O-Ne-Mg WDs. Such WDs are known to be present in some nova systems, possibly as the result of binary evolution.

Although all central stars of PN should evolve to become WDs, not all WDs have necessarily passed through the PN stage. Figure 15.1 shows different pathways to the WD stage. The solid curve represents stars that have evolved through the thermal pulsing and mass loss on the AGB and the evolutionary track from AGB through PN to the WD stage. Even for this group, not all stars will become PN because some may evolve too slowly to illuminate the nebulae before their dispersal (see Section 11.3). The track labeled P-EAGB (post-early AGB) represents stars that do not enter the thermal pulsing AGB stage and probably do not have extensive circumstellar envelopes to be ionized later as PN. The track labeled "AGB-manqué" (failed AGB) represents stars that evolve directly to WDs from the horizontal branch.

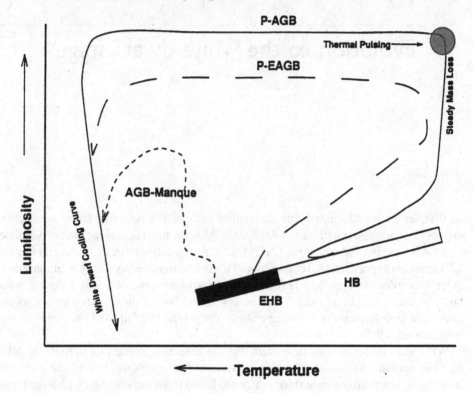

Fig. 15.1. Schematic illustrating different paths to the WD stage (figure from Dorman *et al.*, 1993).

15.1 Structure of white dwarfs

WDs are stars whose state of hydrostatic equilibrium is maintained between a balance of gravitational attraction and electron-degenerate gas pressure. The equation of state is given by $P \propto \rho^{5/3}$ in the non-relativistic case, and $P \propto \rho^{4/3}$ in the relativistic case. In both cases, the pressure is independent of temperature. Combining the equation of hydrostatic equilibrium of a zero-temperature gas with the equation of state gives a mass-radius relationship

$$\frac{R}{R_\odot} = 0.012 \left(\frac{M}{M_\odot} \right)^{-1/3} \left(\frac{\mu_e}{2} \right)^{5/3} \tag{15.1}$$

for the nonrelativistic case, where μ_e is the mean atomic weight per electron. In the relativistic case, there exists a maximum mass (M_∞) that can be supported by electron degenerate pressure:

$$M_\infty = 1.46 \left(\frac{\mu_e}{2} \right)^{-2} M_\odot, \tag{15.2}$$

which is the well-known Chandrasekhar limit.

Figure 15.2 shows the changes in chemical composition from the center to the surface for He and C-O WDs. The masses of the H and He envelopes can be read from the horizontal axes.

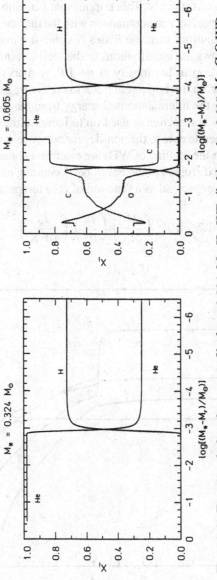

Fig. 15.2. Fractional abundances profile for a 0.324 $M_\odot$ He WD (left) and a 0.605 $M_\odot$ C-O WD (from Blöcker et al., 1997).

15.2   Cooling of white dwarfs

As a result of the combined effects of nuclear burning and mass loss, the H (or He) envelopes of CSPN will eventually be reduced to a point that nuclear reactions are untenable. Energy generation is gradually being taken over by the release of thermal energy by gravitational contraction. On the H-R diagram, the CSPN will turn from the horizontal track to lower luminosities and temperatures. At the termination of H burning, the luminosity of the star drops quickly. This implies that neutrino luminosity from the core will gain increasing importance in comparison with the radiation luminosity. During the early stages of WD evolution, neutrino losses become the main cause for cooling of the core. Figure 15.3 shows the contributions to the total luminosity by the different mechanisms. The sudden drop in luminosity at $t \sim 10^4$ yr marks the extinction of H burning and the downturn from the horizontal track to the cooling track.

A WD cools by releasing the internal thermal energy from the ion gas since the gravitational energy released by contraction is used up to increase the Fermi energy of the electrons through the degenerate core to the non-degenerate envelope. The heat transport mechanisms responsible for the cooling of WDs are electron conduction in the electron-degenerate core and photon diffusion in the envelope. Assuming an isothermal core, the luminosity of a WD can be expressed as a function of core temperature (T_c)

$$\frac{L}{L_\odot} = 1.7 \times 10^{-3} \left(\frac{M}{M_\odot} \right) \left(\frac{\mu}{\mu_e^2} \right) \left(\frac{T_c}{10^7 \, \mathrm{K}} \right)^{3.5}, \tag{15.3}$$

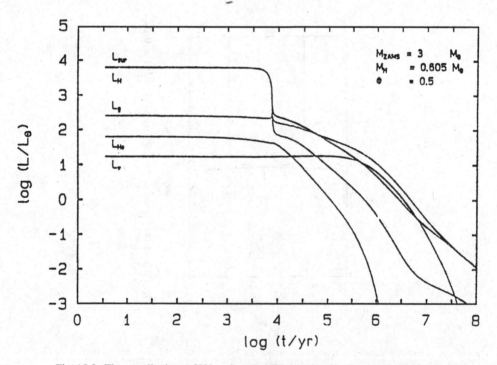

Fig. 15.3. The contributions of H burning (L_H), He burning (L_{He}), gravothermal energy (L_g), and neutrino losses (L_ν) as a function of time for a CSPN of 0.605 $M_\odot$ (initial mass 3 $M_\odot$; from Blöcker, 1995b).

The corresponding cooling time to reach a certain luminosity is given by (Kawaler, 1996)

$$t_{cool} = 9.41 \times 10^6 \, yr \left(\frac{A}{12}\right)^{-1} \left(\frac{\mu_e}{2}\right)^{4/3} \mu^{-2/7} \left(\frac{M}{M_\odot}\right)^{5/7} \left(\frac{L}{L_\odot}\right)^{-5/7}, \tag{15.4}$$

where A is the atomic weight of the ions in the degenerate core and μ is the mean molecular weight of the envelope. For a C-O WD (average A of 14) with a mass of 0.6 $M_\odot$, the cooling time to the faintest WDs observed ($L \sim 10^{-4.5} \, L_\odot$) is $\sim 7 \times 10^9$ yr.

15.3 Envelope chemical composition

WDs are classified spectroscopically according to their optical characteristics. The first letter of the classification (D) indicates a degenerate star, meaning that the lines are sufficiently broad that the surface gravity exceeds $\log g = 7$. The letter D is followed by one or more letters depending on the chemical composition of the atmosphere. For example, DA refers to a WD in which the strongest lines are due to H, DB for a star with HeI lines, DO for a star with HeII lines, DC for a star with a continuous spectrum, DZ for a metallic-line (Mg, Ca, Fe, etc.) star, and DQ for a star with atomic or molecular C features. Additional letters can be added to indicate secondary features (e.g., DAZ). As for MS stars, additional numbers can be added as temperature indicators. For example, DA stars can range from DA1 (hottest) to DA13 (coolest). Sample WD spectra are shown in Fig. 15.4.

The origin of the chemical composition of the atmosphere is uncertain. It is natural to assume that DA WDs descend from H-burning CSPN and non-DA WDs descend from H-poor CSPN (see Section 7.2.1). This interpretation is in general agreement with the fact that approximately 80% of WDs are DA WDs. However, the ratio of observed DA to non-DA is not a constant but changes with temperature. For example, at very high temperatures, very few WDs show measurable H. It is also possible that the surface composition can change during WD evolution as a result of internal processes.

15.4 Subluminous O stars

Hot subluminous subdwarfs (sdO stars) can be found in parts of the H-R diagram occupied by horizontal branch stars, CSPN, and WDs. Since no nebula is found around these stars, they are believed to be either (i) post-AGB stars that evolve too slowly to become PN ("lazy PN"), or post-horizontal branch stars that did not evolve to the AGB. An accurate determination of the space density of sdO stars is necessary to derive the relative contribution to the WD population from different channels (see Section 15.10).

15.5 Transition objects between PN central stars and WDs

PG 1159 stars are H-deficient stars dominated by broad lines of HeII and CIV and by a characteristic absorption caused by HeII λ 4686 Å and various CIV lines (Wesemael *et al.*, 1985). A comparison between the spectra of CSPN with PG 1159 characteristics and those of H-rich CSPN is shown in Fig. 15.5. The prototype, PG 1159–035 (GW Vir), is a pulsating variable that has low-amplitude, nonradial modes with periods between 100 and 1,000 s (see Section 17.5). A non-LTE model atmosphere analysis by Werner and Heber (1991) shows that its effective temperature is $\sim$140,000 K. The effective temperatures of PG 1159 stars range from 75,000 to 160,000 K.

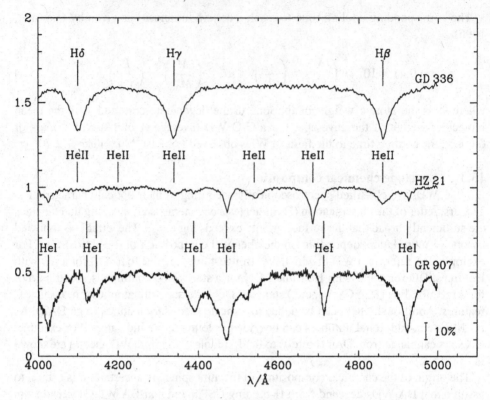

Fig. 15.4. Sample WD spectra illustrating the WD classification scheme. From top to bottom: DA, DO, and DB (from Napiwotzki, 1998).

Although only nine out of ~20 PG 1159 stars are associated with PN, their spectral similarities with WR central stars suggest that these peculiar stars are pre-WDs (see Section 7.2). The evolutionary connection between [WC] stars and PG 1159 stars is very strong in view of their similar C overabundances. An evolutionary sequence from [WCL] (all [WC] stars with subtypes later than [WC6]) to [WCE] (subtypes earlier than WC6) to PG 1159 is suggested by Hamann (1996):

$$[WCL] \rightarrow [WCE] \rightarrow [WC] \rightarrow PG\ 1159. \tag{15.5}$$

As H-deficient objects, PG 1159 stars probably constitute a minority (10-20%) of the transition objects. The main channel of evolution from CSPN to DA WDs is through DAO stars ("O" for low g; see Section 7.2). Like PG 1159 stars, DAO stars are found as central stars of old PN (Napiwotzki and Schönberner, 1995) although some are found to have no PN association (Heber *et al.*, 1996). The comparison between the distributions of CSPN and WD to theoretical evolutionary tracks is shown in Fig. 15.6.

15.6   Luminosity function of white dwarfs

From Eq. (15.4) we see that the cooling time increases sharply as the luminosity decreases. For the faintest WDs, the cooling time is comparable to the age of the universe. Since cool WDs are the oldest stars in the disk of the Galaxy, an accurate luminosity

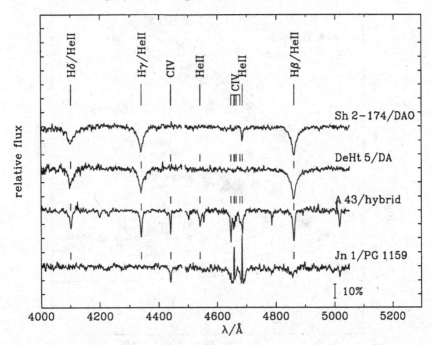

Fig. 15.5. Comparison of the spectrum of a PG 1159 CSPN (Jn 1) with DAO, DA, and a hybrid PG 1159 star (A 43, figure from Napiwotzki, 1998).

function can provide constraints on the initial mass function (IMF) and the star formation rate (SFR) of the Galaxy (see Section 18.2).

Figure 15.7 shows the luminosity function of WDs. The fact that the number of WDs increases with decreasing luminosity can be explained by the increasing cooling time as a WD fades. The cutoff in the low luminosity end is due to the finite age of the Galaxy. By finding very faint WDs in clusters (Fig. 15.8) and determining the cutoff of the luminosity function, one can derive lower limits to the age of the universe (Richer *et al.*, 1998).

15.7   Mass distribution

The observed mass distribution of WDs is similar to that of CSPN in that it is very narrow, with a mean mass of ~0.56 $M_\odot$ and a width of ~0.14 $M_\odot$ (Weidemann, 1990; Bergeron *et al.*, 1992). The narrowness of the mass distribution can be attributed to the increasing mass-loss rate of AGB stars at high luminosities (see Section 10.5). Stars with high initial masses will evolve to higher core masses and therefore high luminosities [cf. the core mass – luminosity relationship, Eq. (10.1)]. They develop very high mass-loss rates, which rapidly deplete the envelope mass, giving the core little time for further growth. This dependence of mass-loss rate with luminosity greatly compresses the mass distribution.

15.8   White dwarfs in external galaxies

The HST has opened the possibility of detecting WDs in external galaxies. A WD candidate has been discovered in NGC 1818, a young star cluster in the LMC. The

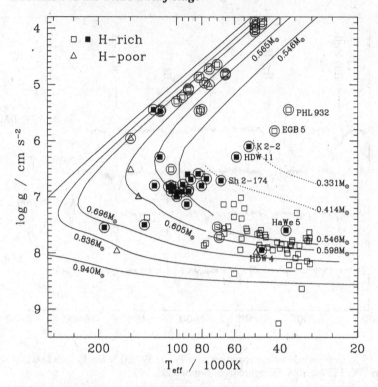

Fig. 15.6. Evolution from CSPN to WD in the log g − log T_* diagram. The squares are H-rich objects and the triangles are H-poor objects. The solid lines are evolutionary tracks of CSPN of different masses. Figure from Napiwotzki (1998).

age of the cluster (4×10^7 yr) implies that the mass of the main-sequence progenitor of this WD is $\geq 7.6\ M_\odot$ (Elson *et al.*, 1998). This suggests that the upper mass limit for the formation of WD in the LMC is similar to that in the Galaxy (see Section 10.7).

15.9   White dwarfs as contributors to dark matter

Stellar remnants make up a fraction of the invisible mass of the Galaxy. The contribution to dark matter by cooled-down WDs in the Galaxy can be calculated by assuming certain galactic evolution models. The derived space densities are dependent on the SFR, the IMF, initial mass-final mass relationship, and the age of the galactic disk (see Section 18.2). Since the upper mass limit for WD progenitors is possibly dependent on metallicity, which changes with time and space through the Galaxy (see Section 18.3), the total number of cooled-down WDs is highly uncertain. Recent estimates have put the WD contribution at ∼4% of the dynamical mass of the Galaxy (Weidemann, 1990).

15.10   Birth rate of white dwarfs

Because of their intrinsic faintness, WDs are known only in the immediate neighborhood of the Sun, mainly within 100 pc. If a complete count to a certain absolute magnitude can be made within a certain volume, then the birth rate of WDs can be derived if the cooling time to that magnitude is known. Weidemann (1990) obtains a birth rate of

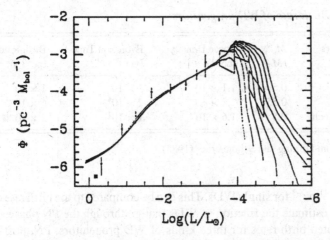

Fig. 15.7. The observed luminosity function of WDs fitted by theoretical curves corresponding to disk ages from 6 Gyr (at the highest luminosity) to 13 Gyr (from Wood, 1992).

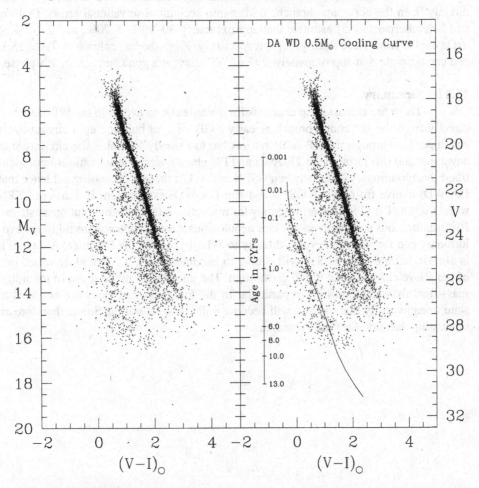

Fig. 15.8. Color-magnitude diagram for the globular cluster M4. The oldest WDs in this cluster are ~9 Gyr (from Richer *et al.*, 1997).

Table 15.1. *Birth rate of WD progenitors*

WD progenitors	M_c $(M_\odot)$	Space Density (pc^{-3})	Evolution Time (yr)	Birth Rate $(pc^{-3}\ yr^{-1})$
PN	0.58	1.1×10^{-8}	1×10^{4}	1×10^{-12}
lazy PN	0.55	1.2×10^{-7}	2×10^{5}	6×10^{-13}
Horizontal branch	0.5	1.6×10^{-7}	1×10^{7}	2×10^{-14}

Table taken from Drilling and Schönberner (1985).

$(0.9 - 1.4) \times 10^{-12}$ yr^{-1} pc^{-3} for single WDs. This can be compared to the birth rate of PN (see Section 18.2) to estimate the fraction of WDs that go through the PN phase. Table 15.1 gives the estimated birth rates for three kinds of WD progenitors: PN, post-AGB stars that do not evolve fast enough to ionize the nebulae (lazy PN), and stars that are not massive enough ($M_c \sim 0.5\ M_\odot$) to ascend the AGB and evolve to the WD stage directly from the horizontal branch. Taking into account observational errors, Drilling and Schönberner (1985) estimate that approximately 30–90%, 5–70%, and 0.2–3% of WDs evolve from these three groups, respectively. From the last column of Table 15.1, one can estimate that approximately 40% of WDs have not gone through the PN phase.

15.11 Summary

There are at least three channels for a single star to evolve to the WD stage: (i) directly from the horizontal branch or early AGB without building up a circumstellar envelope; (ii) through the TP-AGB but evolve too slowly to ionize the circumstellar envelope; and (iii) through the TP-AGB and PN phases. Recent observations have identified the evolutionary links between CSPN and WDs; there is increasing evidence that DA WDs evolve from H-rich CSPN and non-DA WDs evolve from H-deficient CSPN with DAO and PG 1159 stars representing the respective intermediate evolutionary stages. Hot subluminous stars with no obvious nebulosities have the potential to fill in the evolutionary gap for stars that evolve directly to WDs bypassing the PN stage. Although it is almost definite that the Sun will evolve to become a WD, it is not clear which one of the above channels that it will go through. The uncertainty in the shape of the initial mass-final mass relationship, in particular in the low mass end, does not allow us to state unequivocally that the Sun will become a PN. The fate of the Sun is therefore an interesting question for future research.

16

Distances to planetary nebulae

Stars on the main sequence (MS) obey a well-defined luminosity-spectral type relationship. The method of "spectroscopic parallax" assumes that a star of a certain spectral type will have a certain intrinsic luminosity, and by comparing with its visual magnitude, it is possible to derive its distance. However, the central stars of many PN cannot be observed, and even in the case where a visual magnitude is available, their high temperatures imply that large bolometric corrections are required. Whereas stars on the MS remain stationary in the same position on the H-R diagram, CSPN undergo large changes in temperature and luminosity over their lifetime. As a result, any derivations of distances from stellar properties are necessarily model dependent.

Except for a few cases of nearby PN where trigonometric parallaxes are possible, or in cases where there is a MS binary companion, most of the distances to PN have to be estimated from their nebular properties. By making certain assumptions on the nebular structure, distances can be derived by measurements of fluxes, angular sizes, electron densities, and so on. These methods are collectively called statistical distances.

A well-determined distance scale for PN is necessary for the investigation of the space density, galactic distribution, total number of PN, and the birth rate of PN in the Galaxy (see Chapter 18). Unfortunately, after many years of efforts, distances to PN remain controversial. For example, the distances to the well-studied PN NGC 7027 estimated by two different papers published in the same year differ by almost a factor of 10 (178 pc by Daub, 1982, and 1,500 pc by Pottasch *et al.*, 1982). Unless the distances to a large number of PN can be improved, the total number of PN in the Galaxy will remain uncertain by a factor of 2–3.

16.1   Statistical distances

Statistical distances rely on PN's sharing certain common physical properties and the distances are derived from the measurement of two or more nebular parameters. Since individual distances are possible only for a small number of PN, the distances to the majority of PN are found by a variety of statistical methods.

16.1.1   Distance by flux and angular size

The most widely used method of distance determination is the Shklovsky method. Shklovsky (1956a) assumes that all PN have the same mass and that the distance can be derived by observing the nebular flux (e.g., f-f continuum flux F_ν or Hβ flux) and the angular size.

The ionized mass of a PN is related to its electron density in the following way:

$$M_i = \frac{4\pi}{3} n_e \mu_e m_H \epsilon R_i^3, \tag{16.1}$$

where μ_e is the mean atomic weight per electron [(Eq. (4.16)], and ϵ is the filling factor used to allow for the nonspherical geometry of the nebula. The electron density can be expressed in terms of the 5-GHz flux through Eqs. (4.18) and (4.19). Assuming $y = 0.11$, $y' = 0.5$, and $T_e = 10^4$ K, we have

$$n_e = 2.8 \times 10^{19} F_{5\,GHz}^{1/2} \epsilon^{-1/2} R_i^{-3/2} D \text{ cm}^{-3}$$
$$= 4.8 \times 10^3 \left[F_{5\,GHz}(mJy) \right]^{1/2} \epsilon^{-1/2} (\theta/arcsec)^{-3/2} (D/kpc)^{-1/2} \text{cm}^{-3} \tag{16.2}$$

where $\theta = R_i/D$ is the angular radius. Substituting Eq. (16.2) into Eq. (16.1), we have

$$M_i = 2.45 \times 10^{-4} F_{5\,GHz}^{1/2} \epsilon^{1/2} \theta^{3/2} D^{5/2} \text{ g}$$
$$= 7 \times 10^{-5} \left[F_{5\,GHz}(mJy) \right]^{1/2} \epsilon^{1/2} (\theta/arcsec)^{3/2} (D/kpc)^{5/2} M_\odot. \tag{16.3}$$

Therefore distances can be determined by the measurements of flux and angular size:

$$D = 46 \left[F_{5\,GHz}(mJy) \right]^{-1/5} \epsilon^{-1/5} (\theta/arcsec)^{-3/5} (M_i/M_\odot)^{2/5} \text{ kpc}, \tag{16.4}$$

and the derived distances are referred to as Shklovsky distances. This method was in fact first used by Minkowski and Aller (1954), before Shklovsky, to set limits on the distance to NGC 3587. Many different distance scales have been published employing the Shklovsky method (e.g. Cahn and Kaler, 1971; Milne and Aller, 1975; Acker, 1978).

Such distance scales are suspect as we now know that the ionized masses of PN are not constant (see Section 4.5). However, the Shklovsky method can be generalized by assuming there is a mass-radius relationship. Using Eq. (16.3), we find that Eq. (4.39) becomes

$$F_\nu \propto \theta^{2\beta-3} D^{2\beta-5}. \tag{16.5}$$

For example, if one assumes that stellar Lyman continuum photon output is a constant, then Eq. (2.38) implies that $n_e^2 R_i^3$ is a constant. Since $M_i \propto n_e R_i^3$ [Eq. (16.1)], we have $\beta = 3/2$ and from Eq. (16.5) $F_\nu \propto D^{-2}$ (Milne, 1982). Statistical distance scales have been created using certain PN as calibrators to derive the value of β ($\beta = 1$, Maciel and Pottasch, 1980; $\beta = 5/3$, Daub, 1982; Cahn *et al.*, 1992). However, if $\beta = 5/2$, the dependence on distances in Eq. (16.5) disappears altogether. This suggests that the errors in the distances will increase rapidly as β approaches 5/2.

During the lifetime of a PN, the central star temperature changes by a large factor and the number of ionizing photons emitted changes by several orders of magnitude. This will have significant effects on the ionization structure and therefore the value of β. An alternative approach to estimate the value of β is to employ a theoretical central-star evolution model and calculate the change in the optical thickness of the nebula with time (Kwok, 1985b; Marten and Schönberner, 1991). Figure 16.1 shows

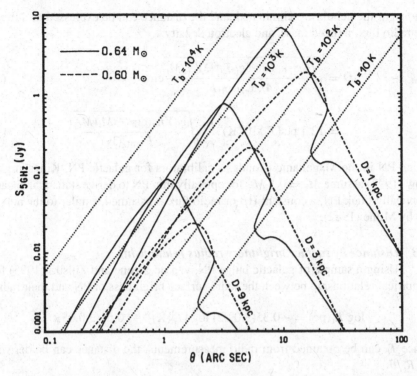

Fig. 16.1. Evolution of the radio flux densities as a function of angular radius for PN with
CSPN of 0.64 $M_\odot$ (solid curve) and 0.60 $M_\odot$ (dash curve) at distances of 1, 3, and 9 kpc. Lines
of constant radio brightness temperatures are plotted as dotted lines (from Kwok, 1985b).

the evolutionary tracks for stars of 0.60 and 0.64 $M_\odot$ in the $F_\nu - \theta$ plane using a simple nebular expansion model and the CSPN evolutionary tracks of Schönberner (1983).
During the ionization-bounded phase (the rising portion of the curves), the tracks of
different distances practically coincide. This implies that a nearby, young PN will look
almost the same as a distant evolved nebula and therefore no distance information can be
extracted. This shows that statistical distance scales are of dubious value for ionization-
bounded PN.

The problem of statistical distances is also illustrated in the study of galactic bulge PN.
Whereas these PN can be assumed to be at the same distance, the statistical distances de-
rived for galactic bulge PN show a large variation of distances, ranging from 4 to >30 kpc
(Stasińska *et al.*, 1991; Pottasch and Zijlstra, 1992).

16.1.2 *Distance by* [OII] *electron density*

Since the optical or radio sizes of PN are sometimes difficult to define, Barlow
(1987) proposes an alternative method using the electron density derived from the [OII]
lines. Since the integrated 3,726/3,729 [OII] doublet ratio is sensitive to density but
only weakly dependent on temperature, it can be used in place of angular size to derive
statistical distances. When Eqs. (4.36) and (16.2) are combined, both the filling factor and

the radius of the nebula are eliminated, and the distance can be expressed as a function of the radio flux, ionized mass, and electron density.

$$D = \sqrt{\frac{6.8 \times 10^{-38} g_{ff} T_e^{-0.5} Y n_e M_i}{4\pi x_e \mu m_H F_\nu}} \, \text{cm},$$ (16.6)

$$= 12.1 \, \text{pc} \, (T_e/10^4 \, \text{K})^{-1/4} \sqrt{\frac{g_{ff} Y n_e (\text{cm}^{-3}) M_i (M_\odot)}{F_{5\,\text{GHz}}(\text{Jy})}}.$$ (16.7)

Using PN in the Magellanic Clouds as calibrators for galactic PN, Kingsburgh and Barlow (1992) assume $M_i = 0.3 \, M_\odot$ for optically thin PN to derive statistical distances. For optically thick PN, a constant Hβ or radio flux is assumed, similar to the approach taken by Milne (1982).

16.1.3   *Distance by surface brightness-radius relationship*
Using a sample of galactic bulge PN, van de Steene and Zijlstra (1995) found an empirical relationship between the radio surface brightness of PN and their radii:

$$\log R_i(\text{pc}) = -0.35(\pm 0.02) \log [T_b(\text{K})] - 0.51(\pm 0.05).$$ (16.8)

Since T_b can be obtained from radio measurements, the distance can be derived by $D = R_i/\theta$.

16.1.4   *Statistical distance scale based on an evolving model of PN*
The main problem inherent in many statistical distance scales is the assumption that one or another physical property of the nebulae is constant. For example, the Shklovsky method assumes that all PN have the same ionized mass. As we have seen, the classical picture of invariant properties of PN is not realistic, as the central star evolves rapidly and is interacting both radiatively and dynamically with the nebula in a time-dependent manner. However, if an empirical formula relating certain nebular properties (e.g., M_i and R_i) can be found, then a statistical distance scale is still possible as long as the power-law index is not too large (see Section 16.1.1).

The observed mass-radius relationship can be explained by a model of nebular expansion coupled with central star evolution (Kwok, 1985b; Schmidt-Voigt and Köppen, 1987a, 1987b). The output of the ionizing photons from the central star first increases sharply with time and then declines as the star moves to the cooling track. Part of the increase of M_i with R_i is due to the snow-plow effect, and part of it is due to the expansion of the ionization front. A number of theoretical $M_i - R_i$ relationships have been constructed based on the H-burning tracks of Schönberner (1983) and Blöcker and Schönberner (1990) and various models of nebular expansion (Volk and Kwok, 1985; Schmidt-Voigt and Köppen, 1987b; Zhang and Kwok, 1993; Mellema, 1994).

Figure 16.2 shows an example of theoretical $M_i - R_i$ curves for different core masses together with the data for 132 PN. An empirical mass-radius relationship of the form

$$\log (M_i/M_\odot) = 1.3077(\pm 0.07) \log (R_i/\text{pc}) + 0.3644 \, (\pm 0.09)$$ (16.9)

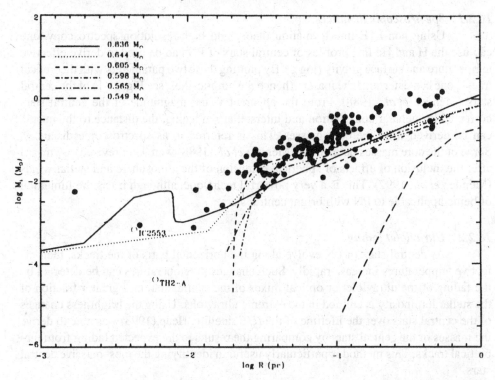

Fig. 16.2. Mass-radius relationship for 132 PN. The thin solid line is the regression line and the other lines represent theoretical curves for different core masses (Fig. 2, Zhang, 1995).

is derived by Zhang (1995). Substituting Eq. (16.9) into Eq. (16.4), and assuming $\epsilon = 0.6$, we have

$$\log D(\text{kpc}) = 1.458 - 0.161 \log(\theta/\text{arcsec}) - 0.419 \log F_{5\,\text{GHz}}(\text{mJy}). \tag{16.10}$$

A catalog of 647 PN distances has been created by using this method (Zhang, 1995).

16.2   Other methods of distance determination

The problems of the statistical methods of distance determination have led to an emphasis on individual distance determinations (Gathier, 1984, 1987). Examples of such methods include the reddening-distance method (Lutz, 1973; Kaler and Lutz, 1985), and the HI 21 cm absorption method (Gathier, Pottasch, and Pei, 1986; Gathier, Pottasch, and Goss, 1986). The observational H-R diagrams derived from these distance estimates still show significant deviations from the theoretical evolutionary tracks (Pottasch, 1989) and are not necessarily more reliable than statistical distances (Zhang, 1993). It is argued by Buckley and Schneider (1995) that the weak dependence of the Shklovsky distance on the ionized mass ($M_i^{0.4}$) lead to errors of $\sim$30% in the derived distance, which is comparable to the errors in individual distance determinations. Below we discuss some of the more recently employed distance-determination techniques.

16.2.1 Spectroscopic distances

Using non-LTE line formation theory and high-resolution spectroscopy, one can use the H and He line profiles of central stars of PN can be used to derive effective temperature and surface gravity ($\log g$). By plotting these two parameters with theoretical tracks, one can determine the masses (hence the luminosities, see Fig. 11.1) of the central stars (Méndez *et al.*, 1988). From the apparent visual magnitude of the central stars corrected for nebular contribution and interstellar reddening, the distance to the nebula can be derived from the core masses. This is referred to as spectroscopic distances. Some of the core masses obtained by Méndez *et al.* (1988) were later revised downward after the inclusion of effects of spherical extension of the atmosphere and stellar winds (Méndez *et al.*, 1992). This is a very powerful technique, although it has the limitation of being applicable to PN with bright central stars.

16.2.2 Ultraviolet fading

As central stars of PN evolve along the horizontal parts of the tracks, their effective temperatures increase rapidly. Such changes in temperatures can be detected by the fading of the ultraviolet or optical fluxes of the central star, as a greater fraction of the stellar luminosity is emitted in the extreme ultraviolet. Using the brightness changes of the central stars over the lifetime of the *IUE* satellite, Heap (1993) was able to derive the masses of the central stars by comparing the results to the expected fading from theoretical tracks. This method is particularly useful in identifying the most massive central stars.

16.2.3 PN with binary central stars

For CSPN that have visual companions, distances can be obtained from the spectral type of the MS companion. This method should yield excellent distances for PN that have visually resolved binary nuclei (Pottasch, 1984).

16.2.4 Trigonometric parallaxes

Trigonometric parallax is the most direct way to measure distance if the object is close enough to the Sun. Recent advances in CCD detectors represent significant improvements over photographic plates in the accuracy of parallax measurements. The use of fainter (and therefore more distant) reference stars also reduces the parallaxes introduced by the reference stars themselves. Parallaxes with an accuracy of 0.5 mas are now possible (see Table 16.1).

16.2.5 Expansion parallax

Another method to determine individual PN distances is by expansion parallaxes. Assuming that the expansion velocity of PN can be obtained from spectroscopic measurements and the tangential expansion of the nebula is the same as the radial expansion, then the distance is given by

$$D = 211 \left[\frac{(V/\text{km sec}^{-1})}{(\dot{\Theta}/\text{mas yr}^{-1})} \right] \text{ pc} \tag{16.11}$$

where V is the radial velocity and $\dot{\Theta}$ is the proper motion.

Table 16.1. *Trigonometric parallexes*

PN	π_{abs} (mas)	Distance (pc)	V	E(B–V)	M_V (1σ Range)
NGC 6720	1.42 ± 0.55	704	15.75	0.08	6.26 (5.20–6.97)
NGC 6853	2.63 ± 0.43	380	14.05	0.09	5.87 (5.48–6.19)
NGC 7293	4.70 ± 0.75	213	13.52	0.03	6.79 (6.41–7.11)
A7	0.17 ± 0.63	≥700	15.48	0.10	≤6.0
A21	1.85 ± 0.51	541	15.96	0.08	7.05 (6.35–7.58)
A24	3.11 ± 1.12	322:	17.37	0.07	9.62 (8.65–10.29)
A29	2.18 ± 1.30	459:	18.31	0.14	9.57 (7.60–10.59)
A31	4.75 ± 1.61	211:	15.52	0.07	8.69 (7.79–9.32)
A74	1.33 ± 0.63	752	17.05	0.12	7.29 (5.90–8.14)
S216	7.67 ± 0.50	130	12.62	0.08	6.79 (6.65–6.93)
PW1	2.31 ± 0.38	433	15.53	0.14	6.91 (6.52–7.24)

Table taken from Harris *etal*. (1997).

Table 16.2. *Planetary nebulae expansion distances*

PN	Θ(''/100 yr)	V_{exp}(km/s)	Distance (pc)	Method
NGC 246	1.4 ± 0.5	38	570	Optical
NGC 2392	0.72 ± 0.06	55	1,600	Optical
NGC 3242	0.83 ± 0.25	20	510	Optical
	1.32 ± 0.47	26 ± 4	420	Radio
NGC 6210	0.31 ± 0.04	23 ± 5	1,570	Radio
NGC 6302	0.18 ± 0.03	13	1,600	Radio
NGC 6572	0.81 ± 0.10	17	440	Optical
	0.25 ± 0.08	14 ± 4	1,180	Radio
	0.25 ± 0.06	14 ± 4	1,200	Radio
BD 30-3639	0.173 ± 0.051	22 ± 1.5	2,680	Radio
	0.30 ± 0.05	22 ± 4	1,500	Radio
NGC 7009	0.70 ± 0.3	21	600	Optical
NGC 7027	0.47 ± 0.7	21	940	Radio
	0.42 ± 0.06	17.5 ± 1.5	880	Radio
	0.525 ± 0.1	17.5 ± 1.5	700	Radio
NGC 7662	0.56 ± 0.15	21	790	Optical
	1.0 ± 0.6	26	550	Optical
	0.26 ± 0.09	21	1,700	Optical
	0.56 ± 0.5	21	790	Radio

Table taken from Terzian (1997).

First efforts in expansion distances were made in the optical. Liller *et al.* (1966) were able to measure the expansions of four PN by using photographic plates taken over time intervals of 44 to 62 yr. High-resolution and dynamic range VLA maps over a time span of several years have yielded expansion distances for several PN (see Table 16.2, Masson 1989a, 1989b; Hajian *et al.*, 1993). In general, the radio expansion distances are greater than those derived with statistical (Shklovsky) methods.

However, even the expansion method is not without pitfalls, as both the expansion velocity inferred from line profiles and image sizes are affected by the dynamical and ionization evolution of the nebulae (see Section 13.5).

16.2.6 Time-correlation method

This method is proposed by Hajian *et al.* (1997) and is based on the assumption that the dynamical time between multiple shells should be equal to the interpulse period. With the use of the observed angular sizes and expansion velocities of the shell and the halo, the dynamical time is derived and compared to the empirical interpulse formula of Frank *et al.* (1994):

$$\log \tau_{ip} = 5.19 - \frac{L_*}{1.78 \times 10^4 \, L_\odot}, \tag{16.12}$$

Since both the dynamical time and L_* are functions of distance, forcing the interpulse time and the dynamical time to be equal will yield a distance. However, as the morphological structures of PN are extensively modified by dynamical effects, it is unlikely that the envelope structures from the AGB could be maintained through the PN phase (see Section 13.6), bringing into question the very premise of this method.

16.3 Summary

Although PN distance is the fundamental parameter for the determination of the luminosities of CSPN and the galactic population of PN, the current state of galactic PN distances remains unsatisfactory. In hindsight, one may wonder why the Shklovsky method, given its obvious limitations, was trusted and in use for so long. The most reliable methods, trigonometric and spectroscopic parallaxes, are applicable to too few PN. The time-correlation method, given its interpretation of the multiple shells, is on too uncertain theoretical grounds to be taken seriously. It is unlikely that accurate distances to a large number of galactic PN will become available anytime soon. PN in external galaxies may therefore offer a better hope of determining some of the properties of PN (e.g., luminosity function, galactic distribution, etc.).

17

Comparison between evolutionary models and observations

The traditional way of comparing stellar evolutionary models and observations is through the Herztsprung-Russell (H-R) diagram. However, we should remember the H-R diagram has its origin in color-magnitude diagrams of stellar clusters, where all stars are at the same distance. For galactic and globular clusters, the stars have relatively small bolometric corrections, and therefore the luminosities and temperatures can be easily derived from apparent magnitudes and colors. The plotting of PN on the H-R diagram turns out to be a much more difficult task. In order for luminosities to be obtained, accurate distances and total fluxes are needed. CSPN have very high temperatures, and the conversion from visual magnitudes to bolometric magnitudes requires a large and uncertain correction factor. A large part of the stellar flux is transferred to the nebula, so the total emitted fluxes from the CSPN have to be inferred from the nebular spectrum. The accounting of the UV flux would be incomplete if the nebula is not ionization bounded. Furthermore, PN emit a large fraction of their total fluxes in the far infrared (see Section 6.2), and observations in the optical region alone are insufficient to account for all the emitted fluxes.

In spite of numerous efforts, the determination of distances to galactic PN remains uncertain to a large factor. Since many of the CSPN are not directly observable, their temperatures have to be inferred from the nebular properties. In the case of the Zanstra method, this implies the necessity of assuming that the nebulae are ionization bounded, which they often are not (Méndez et al., 1992). Because of these uncertainties, the testing of evolutionary models by a comparison with observed distributions of PN in the H-R diagram has often led to erroneous conclusions, as the history has aptly demonstrated (Schönberner, 1993).

17.1   Are PN ionization bounded?

The assumption that PN are ionization bounded is key to the derivation of the luminosities and temperatures of CSPN from nebular properties. This assumption is based on the short mean-free paths of the Lyman continuum photon in a dense nebula (see Section 2.1). However, we now know that most of the PN are not spherically symmetric. Even those that appear circular can just be asymmetric shells that are being viewed pole on (see Chapter 8). The large variation of the density between the pole and equator means that the nebula can be ionization bounded in the equatorial directions but unbounded in the polar directions. If this is the case, then most PN are leaking UV fluxes in the polar directions and their luminosities would be underestimated.

Table 17.1. *CSPN properties for PN with parallax distances*

PN	T_* (K)	L_* ($L_\odot$)	M_c ($M_\odot$)	log g
NGC 6853	1.2×10^5	290	0.6	7.03
A21	1.2×10^5	100	0.65	7.50
NGC 6720	1.2×10^5	200	0.6	7.19
NGC 7293	1.1×10^5	95	0.6	7.36
A74	6.5×10^4	13	0.55	7.32

Table taken from Pottasch (1997).

The detection of faint halos by narrow-band imaging (see, e.g., Fig. 13.13) suggests that many nebulae are totally ionized, and the ionization-bounded assumption used to derive central-star temperatures will result in an underestimation of the star temperature. Since not all of the Lyman continuum photons are accounted for, the total flux (and therefore luminosity) is likely to be underestimated also.

17.2 PN with reliable distances

The availability of better distances has allowed the stellar parameters of certain individual PN to be derived with greater confidence. Table 17.1 lists several examples of PN with parallax distances (see Section 16.2.4). After extinction correction, the absolute visual magnitude of the central star can be used together with temperature to obtain the stellar radius. From the temperature and luminosity, the stellar mass can be obtained by comparing the location of the CSPN on the H-R diagram with theoretical evolutionary tracks. Once the mass is known, the surface gravity can be calculated. These values are found to be in good agreement with those obtained by spectroscopic means (see Section 16.2.1).

Even for this group, discrepancies with theoretical prediction still exist. The predicted evolutionary ages based on the derived core masses are generally higher than the dynamical ages for low luminosity CSPN, and lower than the dynamical ages for high-luminosity CSPN (Pottasch, 1997).

17.3 Test of PN evolutionary models by distance-independent parameters

Since there are large observational errors in the determinations of both luminosity and temperatures of CSPN, observational H-R diagrams for PN are difficult to construct. It is therefore fruitful to explore other means to test evolutionary models of PN other than the H-R diagram.

The Strasbourg-ESO catalog of Galactic Planetary Nebulae (Acker *et al.*, 1992) contains a large number of basic observed parameters of PN. These include the following: angular sizes (θ), radio and infrared continuum fluxes (F_{ff} and F_{IR}), optical nebular line fluxes (F_l), expansion velocities (V_{exp}), visual magnitudes of the CSPN (m_V), and so on. Although most of these observable quantities are distance dependent, it is possible to construct distance-independent parameters from these observables. Examples include line ratios (from F_l), surface brightness (from F_{ff} and θ), central star temperatures

(T_*, from m_V and F_l), electron densities (n_e, from F_l), infrared excess (IRE, from F_{ff} and F_{IR}), dust temperatures (from F_{IR}), nebular colors (from F_{IR}), optical thickness (τ) of the nebula, turnover frequencies of the radio continuum (from F_{ff}), pulsation periods of the CSPN (from m_V), and so on. All of these distance-independent parameters change as a function of time and therefore can serve as tools to test the evolutionary models of PN.

The use of most of the distance-independent parameters, however, test the combined evolution of the nebula and the CSPN. Below we will discuss some examples of the employment of these techniques.

17.3.1 Nebular line ratios

Combining an ionization model with dynamical calculations based on the interacting winds model, Schmidt-Voigt and Köppen (1987a, 1987b) were able to calculate the evolution of the HeII/Hβ line ratio as a function of nebular radius. When the theoretical tracks are plotted against the observational data, core masses of the central stars can be estimated. A completely distance-independent approach was used by Szczerba (1990), who plotted the HeII/Hβ ratio against the Hβ/F$_{cont}$(4,861 Å). Since the former ratio traces the nebular evolution and the latter ratio traces the stellar evolution, the tracks are adequately separated to allow the determination of CSPN masses. For example, Szczerba (1990) was able to show that the mass of NGC 2392 is much lower than the value derived by Méndez *et al.* (1988). He also found that the low luminosity PN of Gathier and Pottasch (1989) are not unusual but in fact lie among the normal PN with masses $\sim$0.60 $M_\odot$.

17.3.2 Surface brightness evolution

The compilations of accurate radio fluxes (Aaquist and Kwok, 1990; Zijlstra *et al.*, 1989) and *IRAS* infrared fluxes have allowed the comparison of the radio and infrared surface brightnesses of a large number of PN (Fig. 17.1). Since both the gas and dust components of PN are expanding, a correlation between these two parameters is not entirely unexpected. What is surprising is that such a correlation implies that there is no substantial difference between the evolution of PN with their CSPN of different masses. In other words, it is difficult to separate a young low-mass CSPN from an evolved high-mass CSPN from nebular surface brightness alone. The only definite statement one can make is that the PN of highest surface brightness can only come from CSPN of high mass. Only high-mass CSPN can evolve fast enough to ionize the nebula while it is still dense (Zhang and Kwok, 1993). The observation of high surface brightness PN therefore allows the identification of the high-mass fraction of the CSPN population.

The ideal combination of distance-independent parameters is that one parameter primarily traces the evolution of the central star while the other primarily traces the expansion of the nebula. The combination of T_b and T_* was investigated by Zhang and Kwok (1993). Although T_b is in principle dependent on the ionization structure, it is mostly determined by the nebular expansion, especially during the optically thin phase. T_* is only a function of central star evolution. For PN in the horizontal parts of the tracks, their evolutionary tracks in the $T_b - T_*$ diagram are sufficiently widely separated such that individual CSPN masses can be determined by interpolation between the tracks. Results are found to be in good agreement with the values determined by Méndez *et al.* (1992).

A similar approach is taken by Tylenda and Stasińska (1994), who use the distance-independent nebular parameter $f = (F_V v_{exp}^2)/\theta^2$ to plot against T_* and compare the

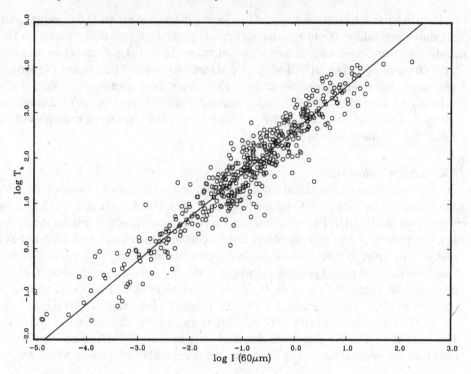

Fig. 17.1. The 5 GHz radio brightness temperature as a function of the 60-μm surface brightness for 476 PN.

resulting data points with the $L_V/(4\pi t_{evo}^2)$ versus T_* tracks from theoretical evolutionary models. They found that although the Schönberner tracks are consistent with observations, the tracks of Vassiliadis and Wood (1994) have too long transition times between the AGB and the PN phase.

17.4   Mass distribution of CSPN

Assuming that the observed properties of PN can be plotted on a diagram where the evolutionary tracks of different CSPN masses are widely separated, for example, on the H-R diagram, the M_V versus $\log t$ diagram of Schönberner (1981), the central star UV mag versus nebular radius diagram of Heap and Augensen (1987), or the T_b versus T_* diagram of Zhang and Kwok (1993), their masses can be derived by comparing their positions with theoretical evolutionary tracks. For example, Schönberner (1981) found the masses of CSPN to have a narrow distribution peaking at ~0.55 $M_\odot$ and Heap and Augensen (1987) found the distribution peaking at ~0.675 $M_\odot$. Figure 17.2 shows the mass distribution of 303 PN. A strong peak centering around 0.60 $M_\odot$ can be seen with a tail in the high mass side extending to >0.8 $M_\odot$.

17.5   Asteroseismology of CSPN

Because of the existence of a well-defined mass-luminosity relationship, the luminosities of CSPN would be known if the stellar mass could be determined by an

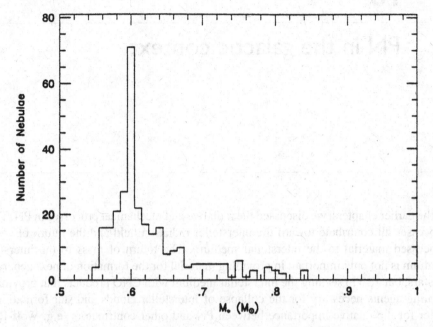

Fig. 17.2. The core-mass distribution of PN (from Zhang and Kwok, 1993).

independent method. One example of such a method is through the pulsation properties of the CSPN.

The first two pulsating CSPN (K1-16 and Lo 4) were discovered by Grauer and Bond (1984) and Bond and Meakes (1990). They are nonradial multiperiodic g-mode pulsators with periods of 25–31 min and low amplitudes (0.05–0.1 mag). Six more pulsating CSPN were found by Ciardullo and Bond (1996). From the pulsation periods measured and the pulsation modes identified, the CSPN masses can be determined by comparison with models. A mass of $0.55 \pm 0.03\ M_\odot$ is found for the central star of NGC 1501 (Bond *et al.*, 1996).

17.6 Summary

In this chapter, we have seen that the traditional method of testing theories of stellar evolution by overlaying the observational distribution of PN on a grid of evolutionary tracks on the H-R diagram has severe limitations. Observationally, the luminosities and temperatures of CSPN are difficult to determine. In order to avoid the most problematic parameter – the distance – a number of ways to test the evolutionary models of PN using distance-independent parameters have been explored.

Unlike stars on the MS, which stay unchanged for billions of years, CSPN undergo significant changes over thousands of years. In this regard, PN share some similarities with WR stars. These changes are manifested in surface properties such as temperature, luminosity, chemical composition, and mass loss. Such rapid evolution of stellar properties makes the evolution of CSPN and that of WR stars two of the most interesting and challenging problems in stellar astrophysics.

18

PN in the galactic context

In the earlier chapters, we discussed the radiative and mechanical processes in PN. These processes all contribute toward the interstellar radiation field and the return of stellar-processed material to the interstellar medium. The return of mass to the interstellar medium is not only important in providing material for the formation of next generation of stars, but also in seeding the interstellar medium with CNO products that provide the cooling agents necessary for the collapse of interstellar clouds and star formation. In order for the relative importance between PN and other contributors (e.g. Wolf-Rayet stars and supernovae) to be assessed, it is necessary to know the total PN population in the Galaxy.

The first estimate of the total population of PN in the Galaxy (6×10^4) was given by Shklovsky (1956b), using the distance scale that he developed. This determination is instrumental in establishing the evolutionary status of PN as a stage that is passed through by all low-mass stars (Abell and Goldreich, 1966). This number also gives us an insight into the star-formation history of the Galaxy, as the progenitor stars of the PN that we are observing today were born billions of years ago.

18.1 Formation rate of PN in the Galaxy

In principle, the derivation of the formation rate of PN (χ) is straightforward once the local number density of PN is known. As a first approximation, we can assume that the number density of local PN (ρ_0) is uniform in the galactic plane and the distribution of PN as a function of perpendicular distance (z) from the galactic plane is given by an exponential law:

$$\rho = \rho_0 e^{-kz}, \tag{18.1}$$

where k^{-1} is the scale height. The density of PN projected onto the galactic plane is therefore:

$$\rho_1 = 2 \times \int_0^\infty \rho \, dz$$

$$= \frac{2\rho_0}{k}. \tag{18.2}$$

The number of PN closer than distance D from an observer at $z = 0$ is

$$N(D) = 2 \times \int_0^D 2\pi p \, dp \int_0^{\sqrt{D^2 - p^2}} \rho(z) \, dz, \tag{18.3}$$

where p is the projected radius from the observer on the galactic plane. Substituting Eq. (18.1) into Eq. (18.3) and integrating, we have

$$N(D) = 2\pi\rho_0 \left[\frac{D^2}{k} - \frac{2}{k^3} + e^{-kD}\left(\frac{2D}{k^2} + \frac{2}{k^3} \right) \right]. \tag{18.4}$$

Assuming that the distances are known, the scale height (k^{-1}) can be obtained from a sample of local PN. If the sample is complete to a certain distance, then the value of ρ_0 can be derived from Eq. (18.4). Using different distance scales, various authors have obtained different values of ρ_0, ranging from 40 to 54 kpc^{-3} (Cahn and Kaler, 1971), 80 ± 15 kpc^{-3} (Cahn and Wyatt, 1976), 55 kpc^{-3} (Daub, 1982), 117 ± 49 kpc^{-3} (Amnuel *et al.*, 1984), and 326 kpc^{-3} (Ishida and Weinberger, 1987). It is clear that the value of ρ_0 is still highly uncertain.

The local birth rate can be expressed as

$$\chi = \frac{\rho_0 V_s}{R_2 - R_1}$$

$$= 1.02 \times 10^{-6} \left(\frac{\rho_0}{\text{kpc}^{-3}} \right) \left(\frac{V_s}{\text{km s}^{-1}} \right) \left(\frac{pc}{R_2 - R_1} \right) \text{kpc}^{-3} \text{ yr}^{-1}, \tag{18.5}$$

where R_1 and R_2 are the minimum and maximum observed radii of PN. For example, if we assume $R_1 = 0.08$ pc, $R_2 = 0.4$ pc, and $V_s = 20$ km s^{-1}, then the Cahn and Wyatt ρ_0 value of 80 ± 15 kpc^{-3} gives $\chi = (5.1 \pm 1.0) \times 10^{-3}$ kpc^{-3} yr^{-1}. From Eq. (18.2), $\rho_1 = 20 \pm 4$ kpc^{-2} if $k^{-1} = 125$ pc. Assuming further that ρ_0 is uniform throughout the Galaxy, then the total population of PN can be found by Eq. (18.4). If the disk extends to 15 kpc from the center of the Galaxy, the total PN population is 14,000. Dividing this number by the dynamical lifetime of PN (t_{dyn}),

$$t_{\text{dyn}} = \frac{R_2 - R_1}{V_s}, \tag{18.6}$$

we have a PN formation rate of ~ 1 yr^{-1}.

The number of galactic PN cataloged is therefore $\sim 10\%$ of the estimated total population. This incompleteness is in part due to the heavy dust obscuration in the galactic plane, and in part due to the faintness of old PN. Many compact PN also remain undiscovered because of their stellar appearances.

18.2 The death rate of main-sequence stars

The PN formation rate can be compared to the rate of stars with appropriate masses leaving the MS now. Since the stars are not forming at the same rate now as in the early phases of the Galaxy, the death rate of stars is a function of both the IMF and the SFR. Although the best determination of the IMF in the solar neighborhood (Miller and Scalo, 1979) yields an IMF that is not a simple power law over all stellar masses, here we will adopt the well-known IMF by Salpeter (1955) for its simplicity:

$$\psi(m) \propto m^{-2.35}. \tag{18.7}$$

Assuming that the IMF is constant over time, and m_u and m_l represent respectively the upper and lower MS mass limits of PN progenitors,

$$\psi(m) = \frac{1.35\,m^{-2.35}}{m_l^{-1.35} - m_u^{-1.35}}. \tag{18.8}$$

The lifetime of a star of mass m on the MS can be approximated by

$$\tau(m) = 10^{10}\,\text{yr}(m/M_\odot)^{-3}. \tag{18.9}$$

If we assume that $m_l = 1\,M_\odot$, corresponding to the lowest mass stars that have time to evolve beyond the MS for a galaxy of age 10 Gyr, then the fraction of stars that will evolve to WDs would be 94% and 96% for $m_u = 8$ and $10\,M_\odot$ (see Section 10.7), respectively. In other words, because of mass loss on the AGB, the great majority of stars will evolve to WDs instead of neutron stars or black holes.

The SFR of galaxies of different types are very different. For example, irregulars have a high SFR, whereas spirals have a lower SFR and ellipticals show little or no star formation. Assuming that the SFR is monotonically decreasing with time, it can be written as

$$F(t) = F(0)e^{-t/\lambda}, \tag{18.10}$$

where t is the time of observation (now) after galaxy formation, $F(0)$ is the SFR at $t = 0$, and λ is a constant that depends on galaxy type. For example, irregulars have high gas contents and have a high rate of star formation and are consistent with having gone through one e-folding of their evolution. Therefore λ can be set to be $\sim$10 Gyr for irregulars (assuming that the age of the universe is $\sim$10–15 Gyr). Spirals, having a small fraction of their mass in gaseous form, are not as active as irregulars in forming stars and can therefore be approximated by $\lambda \sim 5$ Gyr. Similarly, ellipticals can be assumed to have $\lambda \sim 1$ Gyr.

If T denotes the present epoch, then the SFR at which the PN progenitor of mass m were formed is

$$F[T - \tau(m)] = F(0)\exp\left\{-\frac{[T - \tau(m)]}{\lambda}\right\}. \tag{18.11}$$

Combining Eqs. (18.7) and (18.11) and integrating over the appropriate mass range, we find that the number of stars (N) leaving the MS now is

$$N = F(0)\int_{m_l}^{m_u} \psi(m)\exp\left\{-\frac{[T - \tau(m)]}{\lambda}\right\}dm. \tag{18.12}$$

Figure 18.1 shows the stellar death rate as a function of λ for different values of m_l. Because of the shape of the IMF, N is not sensitive to m_u. Since the existence of PN relies on the ionization of the nebula (see Section 11.3), the actual PN formation rate is smaller than N. If we take the current PN formation rate in the Galaxy to be 1 yr^{-1} and assume that 40% of stars that leave the MS do not become PN (see Section 15.10), then $N \sim 1.7\,\text{yr}^{-1}$. From Fig. 18.1, we estimate a value of $F(0)$ of $\sim$6 yr^{-1} (using $\lambda = 5$ Gyr) for the Galaxy.

PN of different masses evolve at very different rates; the number of observable (luminous) PN is dependent on the initial mass-final mass relationship. Figure 18.2 shows

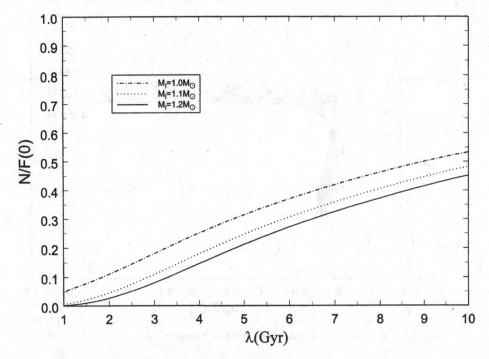

Fig. 18.1. The stellar death rate at $T = 10$ Gyr for galaxies of different star formation constant λ. The upper mass limit is assumed to be 10 $M_\odot$.

a simulation of the distribution of PN in the H-R diagram. A total of 8,355 stars with masses between 0.95 and 10 $M_\odot$ subject to the IMF are randomly generated. The final masses are calculated by using the initial mass-final mass relationship shown in Fig. 10.6. The luminosities and temperatures of the PN are calculated by using theoretical tracks interpolated from the Schönberner models. The ages of the PN are randomly selected between 0 and 40,000 yr after $T_* = 30,000$ K. We can see that all the higher mass PN are on the cooling track. In fact, the least luminous CSPN correspond to a core mass of $\sim$0.65 $M_\odot$ (Blöcker and Schönberner, 1990).

18.3   Effects of metallicity on the formation of planetary nebulae

We now know that PN have their origins in the mass loss from AGB stars (Chapter 12). However, our knowledge of AGB mass loss is entirely based on observations of Population I stars and there is very little direct evidence (e.g., from infrared or millimeter-wave observations) of large-scale mass loss from Population II stars. If AGB mass loss is somehow dependent on metallicity, then the formation rate and the mass range of MS stars that will become PN will also be dependent on metallicity. For example, if the rate of mass loss is dependent on grain formation (see Section 10.5), then metallicity will play a role in the mass loss process. Since different galactic systems have different metal contents, it is possible that the PN formation rates are not uniform in all galactic systems. Since galaxies are enriched over time with each generation of stars, it is also possible that the upper limit for intermediate mass stars may be lower during the early epochs of the Galaxy.

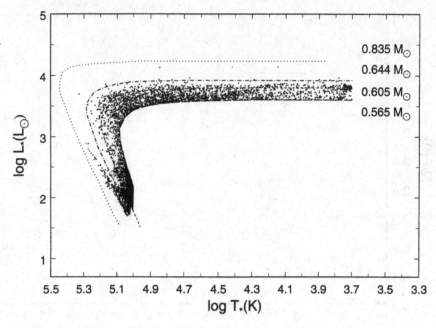

Fig. 18.2. Simulated distribution of PN in the H-R diagram.

18.4   Nearby PN

Nearby PN are expected to have larger angular sizes on the average and can be identified as extended objects. However, the first survey of the POSS plates by Abell (1966) is not complete, and more careful examinations of the plates have revealed many new faint, extended PN. Using a combination of various statistical distance scales, Ishida and Weinberger (1987) cataloged 31 PN within 500 pc from the Sun. From this sample, they derive projected surface density (ρ_1) and space density (ρ_0) of 80 PN kpc^{-2} and 300 PN kpc^{-3}, respectively. Assuming an expansion velocity of 20 km s^{-1}, this yields a local PN birth rate of 8.3×10^{-3} kpc^{-3} yr^{-1}. This is several times higher than the birth rate of WDs (see Section 15.10). If the local sample is representative of the Galaxy, then the total PN population exceeds 10^5.

18.5   PN in the galactic bulge

Are the PN in the galactic bulge different from the rest of PN in the Galaxy? Since PN in the galactic bulge can be considered to be practically at the same distance (approximately 7.8 kpc; Feast, 1987), it is possible to construct a mass distribution and luminosity function for this population of PN. Webster (1988) compared the excitation classes of PN in the galactic bulge and the Magellanic Clouds and found that the bulge PN have lower mass progenitors than the LMC and the solar neighborhood. A sample of galactic bulge PN has been assembled by Stasińska *et al.* (1991) based on the PN spectroscopic survey of Acker and Stenholm (see Section 3.10). With the use of a nebular expansion model and a grid of theoretical evolutionary tracks, the observed Zanstra temperatures and radio/Hβ fluxes are compared with theoretical expectations and the core masses are derived. Figure 18.3 shows the mass distribution of galactic bulge PN.

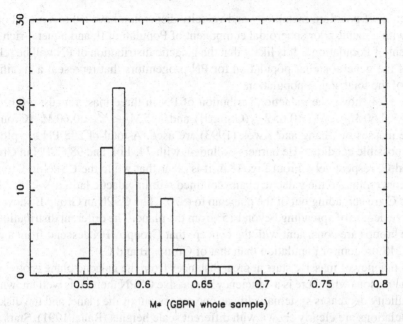

M* (GBPN whole sample)

Fig. 18.3. Histogram of galactic bulge CSPN mass distribution (from Tylenda *et al.*, 1991).

They found a mean CSPN mass of $0.593 \pm 0.025\ M_\odot$. This is slightly lower than the mean value for galactic PN found by Zhang and Kwok (1993).

18.6 PN in the galactic halo

There are approximately 150 globular clusters in the Galaxy, but only four PN are known to exist in globular clusters. They are Ps-1 in M15, GJJC-1 in M22, and two PN (JaFu 1 in Pal 6 and Jafu 2 in NGC 6441) found in the survey by Jacoby & Fullton (cf Jacoby *et al.*, 1997). It is useful to study PN in globular clusters because distances to globular clusters are much better determined than galactic PN, and the PN nebular and central star masses can therefore be estimated with higher accuracy. As in open clusters, the cluster turn-off age gives the minimum initial mass for the PN progenitor, and central star masses measured will give a direct test of the initial mass-final mass relationship.

The rarity of PN in globular clusters can be due to either (i) the central star mass being too low for the star to evolve to high temperature before the dispersal of the nebula (see Section 11.3); or (ii) the low-metallicity AGB progenitors are losing mass at a lower rate resulting in a smaller nebular mass and therefore making the PN more difficult to detect.

It is interesting to note that three of the four PN in globular clusters are found in clusters with an X-ray source. If the X-ray source is associated with interacting binaries, then these PN may also be the result of binary evolution.

18.7 Galactic distribution of PN with different core masses

It is well known that stars of different masses are distributed differently against the galactic latitude. Early-type stars show much higher concentration to the galactic plane, compared to late-type stars. The studies of the stellar distribution in the Galaxy as a function of position, metallicity and velocity dispersion have led to a model of the Galaxy with a metal-poor spheroidal component of Population II, and a metal-rich disk component of Population I. It is likely that the galactic distribution of PN will be related to that of the general stellar population for PN progenitors that represent a significant fraction of the total stellar population.

Figure 18.4 shows the galactic distribution of PN in three mass ranges: $>0.65\ M_\odot$ (Group A), $0.60\ M_\odot < M_c < 0.65\ M_\odot$ (Group B), and $0.55\ M_\odot < M_c < 0.60\ M_\odot$ (Group C). The core masses of Zhang and Kwok (1993) are used. A total of 288 PN are plotted, with the possible candidate He burners excluded, with 74, 114, and 98 CSPN in Groups A, B, and C, respectively. From Fig. 18.4, it is clear that while the CSPN in Group A heavily concentrate to the galactic plane, confined within galactic latitudes $<15°$, those of Group C are spreading out of the plane up to $\approx70°$. The CSPN in Group B show only a moderate degree of spreading beyond 15° from the plane. The different distributions of the three groups are consistent with the concept that Group A PN descend from a more massive MS progenitor population than that of Groups B and C.

The deficiency of massive stars at great distances from the galactic plane is obviously one of the reasons why there is a deficiency in massive CSPN there. It is well known that the metallicity decreases systematically as one moves out of the plane, and the disk and halo populations are clearly shown with different scale heights (Rana, 1991). Stars with greater z distances also show lower metallicity. Papp *et al.* (1983) suggest that a heavy element abundance is a more important factor influencing the formation of PN compared to stellar mass. The deficiency of PN in general at greater z distances indicates that the formation of PN could be hindered by the lack of heavy elements.

18.8   Mass returned to the Galaxy

Mass loss by AGB stars represents the single most important mechanism for mass returned to the interstellar medium. As we have seen, PN are a manifestation of the AGB mass loss and can be considered as part of this process. If we let $\Delta m = M_i - M_c$ represent the total amount of mass returned to the interstellar medium by a star of initial mass M_i, then the total amount of mass (ΔM_{tot}) returned to the interstellar medium at epoch T is given by

$$\Delta M_{\text{tot}} = F(0) \int_{m_l}^{m_u} \Delta(m)\psi(m)\exp\left\{-\frac{[T - \tau(m)]}{\lambda}\right\}dm. \qquad (18.13)$$

For $F(0) = 10$ stars yr^{-1}, $\lambda = 5$ Gyr, $m_l = 1\ M_\odot$, $m_u = 10\ M_\odot$ and assuming the initial mass-final mass relationship in Fig. 10.6, the rate of mass return at $T = 10$ Gyr is 3.3 $M_\odot$.

18.9   Contribution by post-AGB stars to the ultraviolet excess in galaxies

The discovery of UV excess in elliptical galaxies has led to the suggestion that hot post-AGB stars may be responsible (Greggio and Renzini, 1990). However, since PN effectively downgrade most of the UV light into visible light, the observed UV excess can only be explained by naked post-AGB stars, that is, stars without the nebulae. Since stars with low core masses evolve very slowly (see Section 11.1), the nebula would have dissipated when the star became hot enough to emit UV radiation. Figure 18.5 shows the number of PN per unit luminosity as a function of the $1{,}550\ \text{Å} - V$ color of the parent

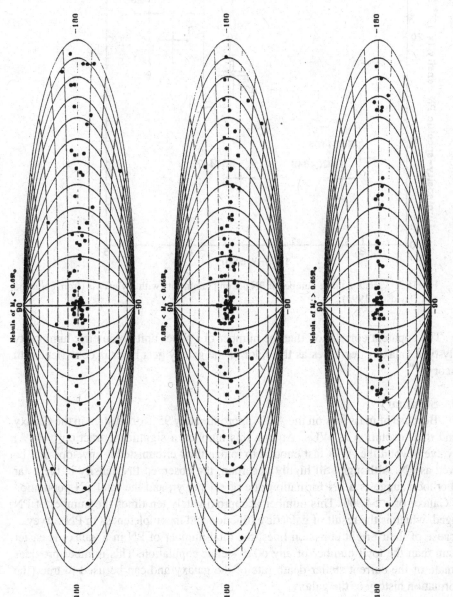

Fig. 18.4. The Aitoff projection maps of PN in galactic coordinates for the central stars with core masses <0.6, 0.6–0.65, and >0.65 $M_\odot$, respectively, from top to bottom panels.

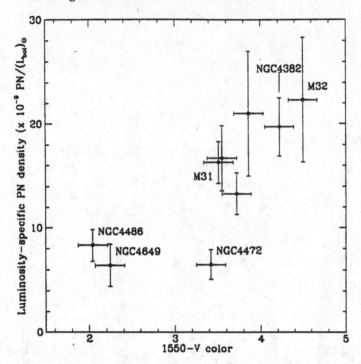

Fig. 18.5. Variation in the number of PN per unit luminosity with UV excess (from Ferguson and Davidsen, 1993).

galaxy. The apparent correlation that the number of PN per unit bolometric luminosity in early-type galaxies decreases as the color of the galaxy gets bluer is consistent with this theory.

18.10 Summary

Because of mass loss on the AGB, approximately 95% of all stars in the Galaxy will end their evolution as WDs. Among this number, a significant fraction (~60%) will evolve beyond the AGB fast enough to ionize their circumstellar envelope and be observed as PN. Although still highly uncertain, the observed PN density in the solar neighborhood suggests a PN formation rate of ~1 per yr, and the total PN population in the Galaxy of ~14,000. This number is approximately ten times the number of PN cataloged, which is the result of galactic extinction and incompleteness in PN surveys.

Because of their bright emission lines, the total number of PN in a galaxy is easier to obtain than the total number of any other stellar population. This number provides a estimate of the current stellar death rate of the galaxy and can be used to trace the star-formation history of the galaxy.

Although PN are bright visual objects caused by nebular line emissions, they do not constitute a significant fraction of the optical brightness of galaxies because of their short lifetimes compared to stars. However, since low-mass post-AGB stars evolve too slowly to ionize their nebulae, this population of post-AGB stars could be a significant source of the UV luminosities of galaxies.

19

Chemical abundances

Progenitors of PN play an important role in the chemical enrichment of the Galaxy because He and significant amounts of C and N are produced by nucleosynthesis during the MS, RG, and AGB stages of their evolution. These elements are brought to the surface of the star by convection events called dredge ups (Iben, 1991). The first dredge up occurs in all stars on the RGB following the exhaustion of H in the core. Convection extending into the interior brings material processed in the CNO cycle to the surface. The second dredge-up occurs only in higher mass ($\geq 3 \, M_\odot$) stars during the early AGB phase and results in an increase in the surface abundance of ^{4}He and ^{14}N at the expense of ^{12}C and ^{16}O. The third dredge-up occurs on the AGB after each He-shell flash, resulting in He, C, and s-process elements being brought to the surface (Renzini and Voli, 1981). This repeated enhancement of C abundance in the stellar surface may in some stars bring the C/O ratio to be greater than unity. Since the less abundant of these two elements is completely tied up in the CO molecule, the photospheric spectrum of the star will change from O-based (e.g., TiO and VO) to C-based (e.g., C_2 and CN), and a carbon star is created (see Section 10.2). Stellar winds during the AGB eject these elements from the stellar surface and therefore enrich the interstellar medium. PN, which are formed from material ejected during the AGB phase, will have their chemical compositions reflect the history of nucleosynthesis in low- and intermediate-mass stars.

19.1   Chemical abundances in planetary nebulae

The abundances in PN trace both the abundances of the interstellar medium from which the progenitor stars were born and the products of nucleosynthesis during the progenitor's evolution. Since He, C, and N are enhanced during stellar evolution but heavier elements such as O and O-burning products and Fe-peak elements are not synthesized in intermediate-mass stars, the study of He, C, and N abundance in PN can test stellar evolutionary models, whereas the abundances of O, Ne, Ar, S, and so on will trace the metallicity of the region where the star was formed. An accurate determination of chemical abundances in PN therefore represents a powerful tool to study not only stellar evolution and nucleosynthesis but also the chemical evolution of galaxies.

PN offer certain advantages over photospheric abundance determination. Because of the presence of strong emission lines, abundances of He, Ne, Ar, and S (which are not usually measured in the spectra of red giants) can be directly measured.

The abundances in PN are determined in the following way. First the measured line strengths are used to derive electron temperature and densities. For example, electron temperatures can be derived from the ratios of [OIII] 4959 Å/4,363 Åand [NII] (6,584 Å + 6,548 Å)/5,755 Å; and electron densities can be derived from the ratios of [SII] 6,717 Å/6,731 Å, [ArIV] 4,711 Å/4, 740 Å, and [CIIII] 5,517 Å/5,537 Å. The abundances of various ions relative to H are then obtained by solving the statistical equilibrium equation using recombination or collisional excitation models. The total elemental abundances are then obtained by applying ionization correction factors, which correct for abundances of unobserved ions. The uncertainties inherent in the last step can be improved by including lines in the ultraviolet. For example, by combining optical and UV lines, one can observe all three ionization stages of carbon.

As we have seen in Section 3.6, the forbidden-line ratios do not necessarily converge to a single point to give a unique solution for T_e and n_e. The fluctuations of either parameter in the nebula can have profound effects on abundance determinations. For example, [OIII] is excited by collisions whereas Hβ is excited by recombination. Since these two processes depend on T_e in different ways, the O^{++} to H^+ abundance ratio determined from the [OIII] to Hβ line ratio can be in serious error.

This problem is a reflection of the fact that PN are dynamical systems in which physical parameters change both in space and time. Accurate chemical abundances can only be calculated with a self-consistent treatment of the dynamical and photoionization processes (Perinotto *et al.*, 1998).

Another major uncertainty is the degree of depletion of atoms in grains. The circumstellar dust of AGB stars contain either O (in silicates) or C (in SiC or amorphous carbon), and the remnants of the AGB dust shell can still be observed in PN. If the dust-to-gas ratio (by mass) is low, then dust depletion will only introduce a small error in the abundance determinations.

The standard method of abundance analysis for He is to use the effective recombination coefficients of Brocklehurst (1971). However, the lowest triplet level ($2s$ ^3S) of HeI is metastable and collisional excitation from this level can excite all the observed HeI "recombination" lines in dense PN (Clegg, 1987).

Pollution from stellar wind is another problem. The abundances of some elements may change during the PN evolution, making the comparison with the theory of nucleosynthesis more difficult. The fast wind from the central star has been invoked to explain the large gas phase depletion of some refractory elements and the large gas phase carbon abundance (Shields, 1980). A WC nucleus with a C/He number ratio of 0.2 and a mass-loss rate of 10^{-8} $M_\odot$ yr^{-1} can inject 3.8×10^{-5} $M_\odot$ of C in 10^4 yr. This is a significant contribution for a nebula of total mass 0.2 $M_\odot$, and a C/H mass ratio of 10^{-3} has only 2×10^{-4} $M_\odot$ of carbon.

19.2   Isotopic abundances

The isotopic abundances of elements are manifestations of the nucleosynthesis and dredge up processes. For example, the $^{12}C/^{13}C$ ratio is predicted to decrease to ~20–30 during the first dredge up and increase again during the third dredge up (Renzini and Voli, 1981). Measurements of this ratio by photospheric (optical) or circumstellar (millimeter-wave) molecular lines have generally yielded values of 7–15 for O-rich AGB stars (Wannier and Sahai, 1987; Smith and Lambert, 1990), although the corresponding values in carbon stars are lower. A measurement of this ratio in the ionized environment of PN was made by Clegg *et al.* (1997) using the $^3P - {}^1S$ multiplet of C^{2+}. While only two lines ($^3P_1 - {}^1S_0$ at $\lambda 1,908.7$ Å and $^3P_2 - {}^1S_0$ at $\lambda 1,906.7$ Å) can be observed in ^{12}C, a third line ($^3P_0 - {}^1S_0$ at $\lambda 1,909.6$ Å) is also allowed in ^{13}C because of its nonzero nuclear spin. Clegg *et al.* (1997) obtained ^{12}C/^{13}C abundance ratios of 15 ± 3 and 21 ± 11 in NGC 3918 and SMC N2, respectively.

19.3   PN abundances in different galactic systems

19.3.1   The galactic disk

Chemical abundances of PN in the galactic disk have been derived by Aller and Czyzak (1983), Aller and Keyes (1987), Henry (1989), Köppen *et al.* (1991) and Kingsburgh and Barlow (1994). A summary of the galactic PN abundance is given by Perinotto (1991). In general, the PN abundances of He, C, and N are enhanced over the values for HII regions. If the HII region abundances are representative of interstellar medium abundances from which stars are formed, then the enhanced abundances of He, C, and N can be attributed to nucleosynthesis and dredge-up episodes.

The mean O/H ratio for PN is $\sim 4.4 \times 10^{-4}$, or 8.6 when expressed as log (X/H) + 12, which agrees with the value in HII regions (see Table 19.1). While most of the surveys concentrate on the most abundant elements such as He, N, O, Ne, S, and Ar, as well as C when UV is observed, lesser abundant elements such as Mg, Al, Si, K, Ca, and Fe have been studied in detail in individual PN (Hyung *et al.*, 1994b; Harrington *et al.*, 1982; Hyung and Aller 1997). The abundance ratios of these elements to O are much lower than solar, which is generally assumed to be due to depletion in grains.

As in HII regions, the O/H abundance ratio in PN is found to decrease from the center to the outer part of the Galaxy. The abundance gradient $[\Delta \log(O/H)/\Delta R]$ has been found to be approximately -0.072 ± 0.012 kpc^{-1} (Faúndez-Abans and Maciel, 1986). The origin of this gradient is not known.

19.3.2   The galactic bulge

The galactic bulge is worth studying because metal abundances can be obtained from both stars and PN. The O abundance in PN are found to be similar to that of bulge stars ([O/H] = log (O/H) − log (O/H)$_\odot \sim -0.3$; Richer *et al.*, 1998). This suggests that PN can be used to trace the O abundances of their progenitor stars.

A large sample of PN in the galactic bulge has been studied by Ratag (1991). Since the galactic bulge PN can be considered to be located at a known distance, the stellar luminosity is less uncertain and therefore would be useful to test any correlations between core masses and abundances. For example, theories suggest that N should be enhanced during the second dredge up, which occurs only in higher mass (between 2.3 and 8 $M_\odot$) stars. Since the O abundance is not changed by AGB evolution, a correlation between the N/O ratio with core mass will confirm the theoretical predictions (see Section 19.4). However, no obvious correlation between N/O ratio and luminosity can be seen in the galactic bulge sample (Pottasch, 1993).

19.3.3   The galactic halo

PN in the galactic halo are defined as high velocity PN at high galactic latitudes. If the halo membership reflects the location where they were born, then the chemical abundances of elements not affected by stellar evolution (e.g., O, Ar, Fe) in these PN will

Table 19.1. *Elemental abundances in different galactic Systems*

	He	C	N	O	Ne	S	Ar
SMC PN	11.03 ± 0.10	8.78 ± 0.42	7.84 ± 0.35	8.24 ± 0.12	7.41 ± 0.18	6.48 ± 0.24	5.62 ± 0.21
SMC HII	10.90 ± 0.02	7.28 ± 0.04	6.46 ± 0.12	8.02 ± 0.08	7.22 ± 0.12	6.49 ± 0.14	5.78 ± 0.12
LMC PN	11.02 ± 0.07	8.74 ± 0.30	8.07 ± 0.34	8.44 ± 0.12	7.70 ± 0.22	6.51 ± 0.20	5.93 ± 0.18
LMC HII	10.93 ± 0.02	7.93 ± 0.15	6.97 ± 0.10	8.43 ± 0.08	7.64 ± 0.10	6.85 ± 0.11	6.20 ± 0.06
Galactic PN	11.00 ± 0.04	8.74 ± 0.21	7.96 ± 0.32	8.68 ± 0.14	7.98 ± 0.20	6.88 ± 0.41	6.31 ± 0.21
Galactic HII	11.00 ± 0.02	8.46 ± 0.20	7.57 ± 0.04	8.70 ± 0.04	7.90 ± 0.10	7.06 ± 0.06	6.42 ± 0.04
Solar	10.99 ± 0.03	8.60 ± 0.05	8.00 ± 0.04	8.93 ± 0.03	8.09 ± 0.10	7.21 ± 0.06	6.56 ± 0.10

Here (log (X/H)+12); table taken from Clegg (1997).

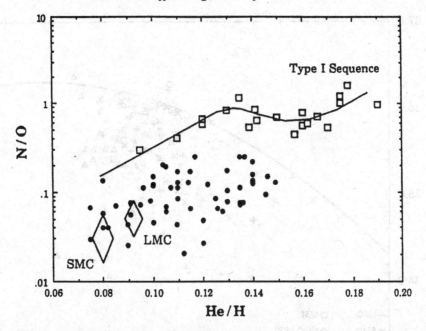

Fig. 19.1. Variation of N/O with He/H for PN in the LMC and SMC. The diamonds represent the mean interstellar medium abundance ratios (from Dopita and Meatheringham, 1991b).

reveal the chemical composition of the halo at the time the stars were born. Studies of the small sample of halo PN suggest that O and Ne are underabundant by 1 order of magnitude whereas Ar, Fe, and S are underabundant by 2 orders of magnitude compared to PN in the solar neighborhood.

19.3.4 The Magellanic clouds

Chemical abundances of PN in the Magellanic Clouds have been studied extensively (Monk *et al.*, 1988; Dopita and Meatheringham, 1991a, 1991b; Clegg, 1997). A summary of PN abundances in the SMC, LMC, and the Galaxy is shown in Table 19.1. For Ne and Ar, elements that are least affected by nucleosynthesis, their abundances are lower in the LMC and the SMC than those in the Galaxy. Whereas the abundances of O and Ne are similar to those found in HII regions in the LMC and SMC, the abundances of N in PN are significantly enhanced over that in HII regions. Given the low heavy abundances of the LMC and SMC, this suggests that almost all of the initial C has been converted to N and brought to the surface during the first dredge up phase.

Figure 19.1 shows that the abundances of N and He in LMC and SMC PN are enhanced over the mean interstellar medium values. A separate Type I (objects with N/O > 0.3) sequence can also be seen.

19.3.5 Other galaxies

Galaxies are chemically enriched by successive star formation, and the evolution of galaxies can be traced by their metallicity. Since PN are found in all types of galaxies, they can be used as probes of chemical abundances in galaxies. They are particularly useful in galaxies where there is no active star formation and therefore have

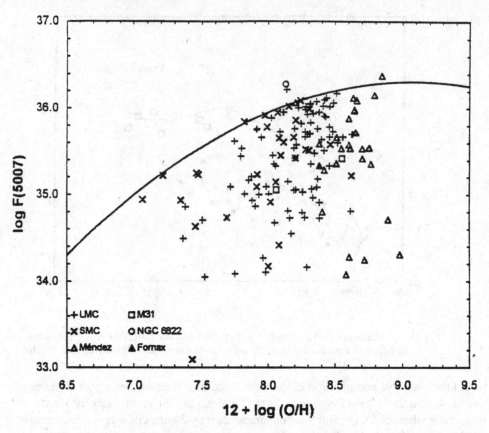

Fig. 19.2. Relation between [OIII] luminosity and O abundance in galaxies. The solid line is Eq. (19.1) (figure from Richer and McCall, 1995).

no HII regions to use for chemical abundance determinations. For example, O is mainly produced by Type II supernovae and its abundance is not expected to be affected by AGB nucleosynthesis; the O abundance obtained from PN should reflect the initial abundance when their progenitors were formed. Based on the observation that the maximum [OIII] λ 5,007 luminosity in Magellanic Cloud PN varies with nebular O abundance, Richer and McCall (1995) suggest that the minimum O abundance can be obtained by measuring the [OIII] flux using the following formula:

$$\log L(5,007)_{max} = 11.46 + 5.47x - 0.301x^2 + 0.4(\mu_{LMC} - 18.371), \qquad (19.1)$$

where $x = 12 + \log(O/H)$ and μ_{LMC} is the distance modulus of the LMC. Since [OIII] is easily measurable (see Chapter 20), this method allows the determination of O abundance in many galaxies.

19.4   The relation between chemical abundances and core masses

It is often assumed that the chemical abundances of PN are dependent on the mass of the central star. Peimbert (1978) classifies PN into four classes. Type I have high He or N abundances, usually show bipolar morphology, and are assumed to have come

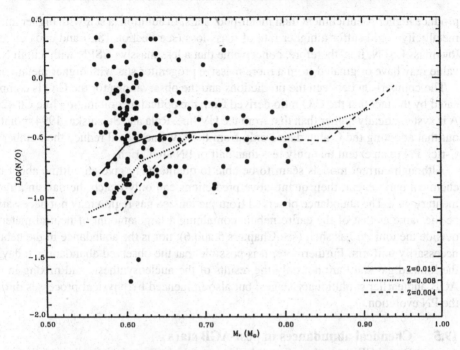

Fig. 19.3. The N/O ratio plotted against the core mass of the central star. Dots are data points. Solid, dotted, and dashed curves are theoretical predictions of Vassiliadis (1992) for three different initial metallicities (Figure from Zhang and Kwok, 1999).

from a high-mass population. Types II and III are galactic disk nebulae with respectively low- and high-velocity dispersions perpendicular to the galactic plane. Type IV are halo objects. This is a sequence of decreasing initial mass and increasing age.

Confrontations of the theory with observations have been carried out by many authors. In particular, Kaler and Jacoby (1990, 1991), Kaler *et al.* (1990), Stasińska and Tylenda (1990) have recently compared the observed abundance ratio N/O, as a function of the core mass of CSPN, with the theoretical predictions of Becker and Iben (1979, 1980) and Renzini and Voli (1981). Although these studies suggest that the observational abundance ratios of PN can be qualitatively explained by the current dredge-up theory, observational scatters around the theoretical predictions have remained large.

Evolutionary models from the MS to the WD stage for stars of mass 1–5 $M_\odot$ and different initial compositions have been computed by Vassiliadis (1992). The surface compositions of He, N, C, and O are also obtained after the third dredge up as a function of initial mass and composition. From these models, it is possible to compare the chemical abundances with the final mass, without invoking an empirical relation between the initial and final mass. An example of such calculations is shown in Fig. 19.3, where log (N/O) is plotted against the masses of CSPN using the core masses from Zhang and Kwok (1993). The theoretical predictions of Vassiliadis (1992) are also plotted for three different initial metallicities of Z = 0.016, 0.008, and 0.004.

The theoretically predicted N/O ratios match better for the lower values of log (N/O) between −1.2 to ≈0.0 but fail to account for the higher values up to log (N/O) ≈ 0.5, especially at lower masses around 0.6 $M_\odot$. The theory suggests that metal-rich stars produce higher N enrichment than metal-poor stars. In addition, a star with higher initial metallicity could suffer a higher rate of mass loss (see Section 18.3) and ends up as a low mass CSPN. It is, therefore, conceivable that a less massive CSPN with a high N/O value may have originated from a more massive progenitor star with higher metallicity.

The comparison between the predictions and the observations for the C/O is complicated by the fact that the C/O ratio derived from the optical recombination line CII 4267 Å is systematically higher than that from the UV line. Rola and Stasińska (1994) pointed out that adopting the C abundance only from the UV line would reduce the number of C-rich PN to an extent that only less than half of PN are C-rich.

Although current models seem to be able to qualitatively give an outline about the chemical enrichment, their quantitative predictions can only match the data in a rudimentary way. The abundance observed from the ionized gas region may not necessarily be the same as that of the entire nebula containing a large amount of neutral material outside the ionized gas shell (see Chapters 5 and 6); nor is the abundance in the nebula necessarily uniform. Furthermore, it is possible that the observed abundances today in the ionized gas shell are not only the results of the nucleosynthesis and mixing in the AGB and earlier evolutionary stages, but also influenced by physical processes during the PN evolution.

19.5   Chemical abundances in post-AGB stars

Post-AGB stars can offer valuable complementary data for the the testing of AGB nucleosynthesis. Many post-AGB stars are bright visual objects, and their photospheric abundances can be determined by optical spectroscopy and classical stellar atmosphere analysis. As a result, one can bypass the uncertainties associated with ionization correction factors (see Section 19.1). PPN also suffer far less extinction than AGB stars and do have the complicating molecular bands in their optical spectra. Most of the PPN that have their abundances determined show elevated C, N, and O abundances in comparison with normal Population I supergiants. This is consistent with the fact that they have gone through the He-burning phases of evolution. Specifically, Decin *et al.* (1998) found that the carbon-rich, 21-μm sources (see Section 14.4.4) are metal deficient (e.e., [Fe/H] $= -0.7$ for IRAS 04296+3429), but have large overabundance of s-process elements ([s/Fe] $= +1.4$ for IRAS 04296 + 3429). Figure 19.4 shows the elemental abundances relative to solar values for IRAS 22223 + 4327.

A number of bright, high-latitude supergiants have been found to have extraordinarily low iron-group abundances. The best examples are HR 4049 ($m_V = 5.47$, B9.5 Ib-II, $b = -57°$) and HD 52961 ($m_V = 8.50$, F8 I, $b = 7.6°$), which have [Fe/H] ratios of -4.4 and -4.5, respectively (Waelkens *et al.*, 1991). However, the abundances of light elements C, N, O, and S are approximately solar. Such low iron abundances cannot be the values the stars originally had when they were on the MS. If this were the case, there should be numerous other metal-deficient red giants near the Sun, which are not observed. Since these stars often have infrared excesses, it is likely that the iron-group elements have been selectively removed from the photosphere into grains.

The possibility also remains that these are not really post-AGB stars (see Section 14.2.1). The fact that all stars showing depletion of refractory elements are binaries suggests that the infrared excesses originate from circumstellar disks (van Winckel, 1999). None of these stars show s-process enhancements as evidence that they have

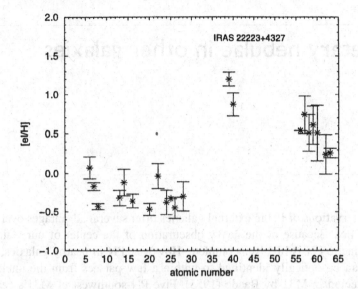

Fig. 19.4. Elemental abundances in IRAS 22223 + 4327 relative to solar values. Note the high abundances of the *s*-process elements: Y(Z = 39), Zr (40), Ba (56), La (57), Ce (58), Pr (59), Nd (60), Sm (62), and Eu (63) (from Decin *et al.*, 1998).

gone through AGB thermal pulses and a third dredge up. Their high galactic latitudes and Population II membership suggest that these are low-mass stars and may have left the AGB before the commencement of thermal pulses.

19.6 Summary

The contribution of stellar nucleosynthesis to the origin of elements is one of the most important questions in astrophysics. The dredge up of newly synthesized elements from the core of AGB stars, and their subsequent ejection from the stellar surface, provide the opportunity to quantitatively test the theories of nucleosynthesis. PN, with easily measured emission lines, play an important role in this process. Although the general picture of the synthesis of He, C, and N in AGB stars is confirmed by observations of PN, many details of the process, for example, whether N is produced in higher quantities in high-mass stars, are still uncertain.

PN also allow us to observe the effects of nucleosynthesis under different initial metallicity conditions. Chemical abundances of PN can be determined in an increasing number of external galaxies and can be used to trace the chemical evolution of galaxies. The difference in mean abundances of PN and HII regions will set constraints on how the metallicity of a galaxy changes with age.

20

Planetary nebulae in other galaxies

Paradoxically, observations of PN in external galaxies offer several advantages over the study of galactic PN. Because of the heavy obscuration of the center of our Galaxy, our catalog of galactic PN is highly incomplete. However, PN in nearby galaxies, for example, M31, can be optically identified to within a few parsecs from the nucleus. PN were first detected in M31 by Baade (1955). Five PN southwest of M31's center were detected in [OIII] with a 1,500-Å filter and the Palomar 5-m telescope. By using a narrower filter (23 Å), Ford and Jacoby (1978) were able to go much fainter and detected 315 PN in the bulge of M31. Later surveys have cataloged PN in M32, NGC 205, NGC 185, and NGC 147.

A survey of PN in an external galaxy can in fact provide a much better picture of the spatial distribution of PN. As discussed in Chapter 16, the distances to galactic PN are highly uncertain. Since the PN in an external galaxy can be safely taken to be at the same distance, the identification of a large number of PN in a galaxy of known distance can lead to an accurate luminosity function.

CSPN are cores of AGB stars and are therefore very luminous objects. If the nebula is ionization bounded, the central star's radiation is reprocessed into line radiation. For example, in the nebular model shown in Fig. 4.3, $\sim$3% of the stellar luminosity (or $\sim$300 $L_\odot$) comes out in the [OIII] line. The observation of a galaxy with a narrow-band 5,007-Å filter can easily distinguish PN from stars.

20.1   Planetary nebulae in the Magellanic Clouds

PN in the Magellanic Clouds have the advantage that their distances are probably known to within 15%, and their distribution in the H-R diagram is reduced to the determination of their total fluxes and effective temperatures. By comparing the observed nebular fluxes to a photoionization model, Dopita and Meatheringham (1991a, 1991b) were able to derive the stellar properties for 82 Magellanic Cloud PN. The observational H-R diagram for the LMC PN are shown in Fig. 20.1.

Figure 20.2 shows the core-mass distribution for 80 Magellanic Cloud PN. The distribution shows a strong peak around 0.6 $M_\odot$ with a tail toward the high-mass end. In order to make a comparison with the selection criteria used for extragalactic distance scale studies, a "bright sample" consisting of only objects within 1.5 mag of the upper [OIII] luminosity limit is constructed by Kaler and Jacoby (1991). They show that the core mass distribution is consistent with a Gaussian distribution with a width of 0.04 $M_\odot$.

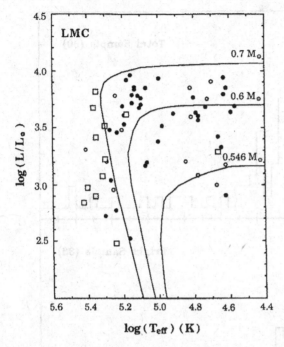

Fig. 20.1. The H-R diagram for 82 MC PN (from Dopita and Meatheringham, 1991b).

20.2   Galactic evolution

The majority of galaxies have no nonthermal activity, and their luminous output is made up primarily of integrated starlight. As a result, the evolution of the SED of galaxies is a reflection of the changing stellar population. Since 95% of stars pass through the post-AGB phase (see Section 18.2), observations of the PN population in galaxies of different redshifts will test the evolutionary model of galaxies. The SED of galaxies can be synthesized by integrating the stellar spectrum as stars evolve (see Section 18.2). For example, as low mass stars leave the AGB, those whose masses are too low to form PN will contribute increasingly to the UV output of the galaxy.

A thorough understanding of PN evolution is necessary for studies of galactic evolution for AGB stars and PN are major contributors to the total bolometric luminosity of a galaxy (Renzini and Buzzoni, 1986).

20.3   Planetary nebulae as distance indicators

Since the discovery of the expansion of the Universe by Hubble, there has been a continuous effort to find the expansion rate of the Universe (and therefore the age of the Universe) by determining an accurate value of the Hubble constant. Since the receding velocities of the galaxies can be measured relatively precisely by the redshifts of spectral lines, the problem lies in distance determination. For nearby galaxies, Cepheid variables have been used successfully to serve as primary distance indicators. For more distance galaxies, brighter objects such as supergiants, novae, or supernovae have been used as secondary distance indicators. The central stars of PN are very luminous objects although their luminosity lies primarily in the UV. During the ionization-bounded stage

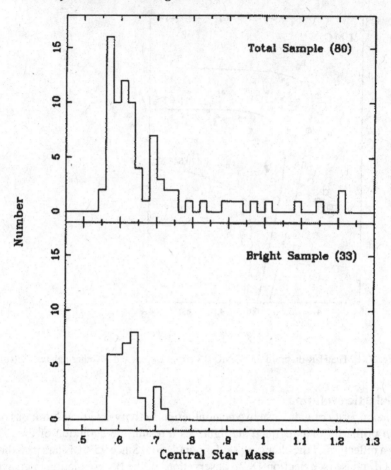

Fig. 20.2. Core-mass distribution for 80 MC PN and for the bright sample (from Kaler and Jacoby, 1991).

of the nebular evolution, most of the UV photons are reprocessed into nebular emission in the visible and can be detected at large distances. Although recent *HST* observations have extended the detection range of Cehpeids (Kennicutt *et al.*, 1995), they cannot be observed in early-type galaxies. PN therefore have the potential to play an important role in the determination of the extragalactic distance scale.

The idea that PN may be useful as "standard candles" for the determination of extragalactic distance scale was first made by Hodge (1966). Noticing that the brightest PN in the local group galaxies have similar [OIII] absolute fluxes, Ford and Jenner (1978) used the brightest PN in M81 and M31 to derive a distance ratio of ~4 between these two galaxies. The serious adoption of the use of PN for distance determination came after the discovery that the planetary nebulae luminosity function (PNLF) has a sharp truncation at the bright end. Assuming that the form of the PNLF is universal, the distance to any galaxy with a large enough number of PN can be obtained by fitting a calibrated PNLF to the observed distribution. By combining the exponential behavior observed in the luminosity function for faint PN and the sharp cutoff observed in the bright end,

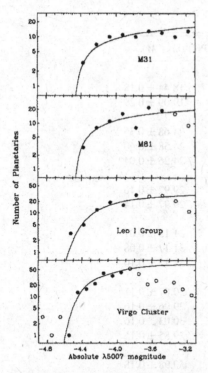

Fig. 20.3. The observed PNLF for M31, M81, three galaxies in the Leo I group, and six galaxies in the Virgo cluster (from Jacoby *et al.*, 1992).

Ciardullo *et al.* (1989) propose the following empirical form for the PNLF:

$$N(M_{5007}) \propto e^{0.307 M_{5007}} \left[1 - e^{3(M_* - M_{5007})} \right] \tag{20.1}$$

where the magnitude M is related to the 5,007-Å flux by

$$M_{5007} = -2.5 \log F_{5007} - 13.74, \tag{20.2}$$

and M_* is the absolute magnitude of the most luminous PN. By adopting the infrared Cepheid distance of 710 kpc and a total foreground extinction of $A_{5007} = 0.39$, Ciardullo *et al.* (1989) found the best-fit value for M_* to be -4.48. With the more recent M31 distance of 770 kpc based on the Cepheid distance scale, the value of M_* is revised to -4.54.

As of 1997, distances to 34 galaxies have been determined by the PNLF method (see Table 20.1). Although the PNLF method was first developed for early-type galaxies, the technique has been extended to spiral galaxies (Feldmeier *et al.*, 1996, see Fig. 20.4). A comparison with other techniques, for example, Cepheid distances, suggests that the PNLF technique is accurate to ~8% (see Fig. 20.5).

20.3.1 Physical basis of the PNLF

The PNLF of a galaxy represents the luminosity distribution of PN in one single instant in time. Since PN of different masses evolve at very different rates and emit greatly varying amount of [OIII] fluxes, the PNLF represents a convolution of central

Table 20.1. *PNLF distances*

Galaxy	Type	Nr. PN	$(m - M)_0$
Local Group			
LMC	SBm	42	18.44 ± 0.18
SMC	Im	8	19.09 ± 0.29
185	dE3p	4	...
205	S0/E5p	12	24.68 ± 0.35
221	E2	9	24.58 ± 0.60
224	Sb	104	24.26 ± 0.04
NGC 1023 Group			
891	Sb	34	29.97 ± 0.16
1023	SB0	97	29.97 ± 0.14
Fornax Cluster			
1316	S0p	58	31.13 ± 0.07
1399	E1	37	31.17 ± 0.08
1404	E2	19	31.15 ± 0.10
Leo I Group			
3377	E6	22	30.07 ± 0.17
3379	E0	45	29.96 ± 0.16
3384	SB0	43	30.03 ± 0.16
3368	Sab	25	29.85 ± 0.15
Virgo Cluster			
4374	E1	37	30.98 ± 0.18
4382	S0	59	30.79 ± 0.17
4406	S0/E3	59	30.98 ± 0.17
4472	E1/S0	26	30.71 ± 0.19
4486	E0	201	30.73 ± 0.19
4649	S0	16	30.76 ± 0.19
Coma I Group			
4278	E1	23	30.04 ± 0.18
4494	E1	101	30.54 ± 0.14
4565	Sb	17	30.12 ± 0.17
NGC 5128 Group			
5102	S0p	19	27.47 ± 0.22
5128	S0p	224	27.73 ± 0.04
5253	Amorph	16	27.80 ± 0.29
Other			
Bulge	Sbc	22	14.54 ± 0.20
300	Sc	10	26.78 ± 0.40
3031	Sb	88	27.72 ± 0.25
3109	Sm	7	26.03 ± 0.30
3115	S0	52	30.11 ± 0.20
4594	Sa	204	29.76 ± 0.13
5194/5	Sbc/SB0	38	29.56 ± 0.15
5457	Sc	27	29.36 ± 0.15

Table taken from Jacoby (1997).

Fig. 20.4. The [OIII] 5007 Å image of the Sc galaxy M101 (NGC 5457) with the positions of the PN marked by crosses (from Feldmeier *et al.*, 1996.)

star evolution, nebular evolution, and mass distribution. The PNLF [N(L$_{OIII}$)] at a given time t can be calculated from L$_{OIII}$(M_c, t) with a central star model [L_*(M_c, t), T_*(M_c, t)] and a nebular model [M$_s$(M_c, t), n_e(M_c, t), and so on]. Figure 20.6 shows the change of the nebular [OIII] fluxes for five central star masses based on the Schönberner tracks. The nebular expansion is given by the momentum-conserving case of the ISW model, and the [OIII] fluxes are calculated using the photoionization code *CLOUDY*. We can see that high mass nebulae contribute a large amount of [OIII] fluxes over a short period of time, and PN with masses lower than 0.565 $M_\odot$ evolve too slowly to emit any significant amount of [OIII].

Figure 20.7 shows the PNLF of Jacoby (1989), assuming a Gaussian distribution of core masses. A more realistic core-mass distribution will require knowledge of the IMF, the SFR in the past, and the initial mass-final mass relationship (see Section 18.2).

20.3.2 *Possible sources of error with the PNLF method*

Since the PNLF represents the numbers of PN in each brightness interval, it is important to have an homogeneous sample. HII regions represent the major source of confusion. For early-type galaxies with few HII regions, this is not a significant problem. However, in late-type spiral and irregular galaxies in which there are more HII regions than PN, some of the [OIII] bright objects can be HII regions. HII regions, being on the

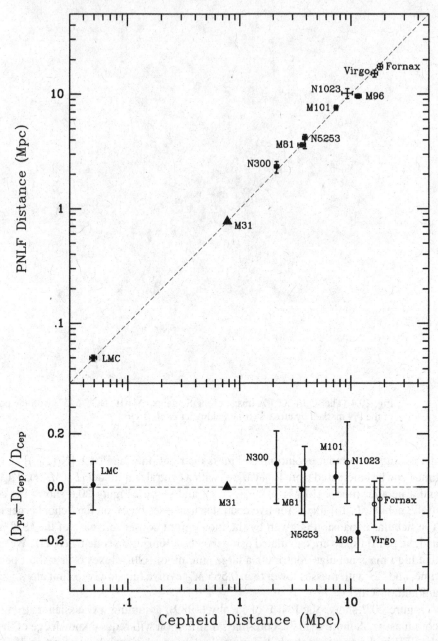

Fig. 20.5. Comparison between PNLF and Cepheid distances. The relative differences between the two methods are shown in the lower panel (from Jacoby, 1997).

average ten times larger in size than typical PN, can be positively separated if they are spatially resolved. A PN of size 0.1 pc will have an angular size of 0.02 arc sec at 1 Mpc and will appear stellar, whereas a HII region can be resolved by a space-based telescope even at distances beyond 10 Mpc.

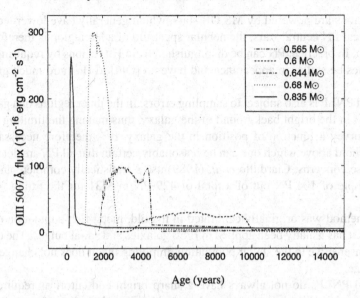

Fig. 20.6. Time evolution of the [OIII] flux for PN with central stars of five different masses (from Tremblay and Kwok, 1997).

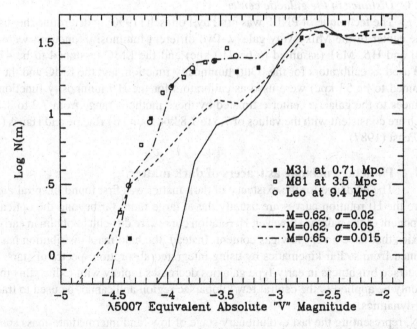

Fig. 20.7. Model PNLF assuming Gaussian distributions of core masses (from Jacoby, 1989).

Since HII regions are powered by MS OB stars, which generally have lower temperatures than typical PN central stars, the nebular spectrum of a HII region is therefore of lower excitation. In practice, PN can be distinguished from HII regions by requiring that all PN candidates be stellar in appearance and have an [OIII] to Hα ratio much greater than one.

The observed PNLF is also subject to sampling errors. In the inner regions of a galaxy, faint PN are lost in the bright background of the galaxy, thus making the limiting magnitude of the survey a function of position in the galaxy. It is therefore necessary to estimate a threshold above which one can be reasonably certain that all PN are detected. Because of these concerns, Ciardullo *et al.* (1989) use a statistically complete and homogeneous sample of 104 PN out of a total of 429 PN in M31 for the fitting by the PNLF.

The PNLF method was originally calibrated in the old, metal-rich population of the M31 bulge, and it has usually been applied in E/S0 galaxies and spiral bulges. The cutoff is found to be insensitive to the stellar population, implying that it does not change with metallicity or age.

The observed PNLFs do not always have a sharp bright-end cutoff as required by Eq. (20.1). In several galactic systems, overluminous objects have been found. Jacoby *et al.* (1996) argue that these objects should be excluded from the sample and suggest several possible origins for their existence. These include HII regions, Wolf-Rayet nebulae, chance superposition of PN, background quasars, coalesced binaries, etc.

20.3.3   Distance to the galactic center

The technique of PNLF was used by Pottasch (1990) to determine the distance to the center of the Milky Way galaxy. Two different luminosity functions were used: [OIII] and Hβ. M31 (assumed to be 710 kpc) and the LMC (assumed to be 47 kpc) were used as calibrators for the [OIII] luminosity function, and the LMC and the SMC (assumed to be 53 kpc) were used as calibrators for the Hβ luminosity function. The distances to the galactic center estimated by these methods range from 7.2 to 8.1 kpc, which are consistent with the values of 7.5 to 7.8 kpc found by Racine and Harris (1989) and Feast (1987).

20.4   Planetary nebulae as tracers of dark matter

The evidence for the existence of dark matter was first found in spiral galaxies where the HI rotation curves are usually flat at large radii, far beyond the optical disk component of the galaxy. However, HI rotation curves are difficult to obtain in early-type galaxies that have a very low gas content. Instead, the halo mass distribution has to be obtained from stellar kinematics by using integrated absorption spectra of stars. Since the surface brightness in early-type galaxies decreases rapidly with radius, this method can only be applied to the central few kiloparsec region and cannot be used to trace the halo dynamics.

PN, representing the last evolutionary stage of low- and intermediate-mass stars, are abundant in an old population, such as in the halo of early-type galaxies. As we have seen, the central stars of PN are very luminous objects. Although they emit primarily in the UV, the nebulae effectively redistribute much of the energy output into the visible in the form of forbidden lines. Because of the high concentration of energy in a few emission lines, PN are easily detectable with narrow-band interference filters. The radial velocities of PN can be efficiently measured with a multifiber spectrograph and therefore can serve as ideal candidates to trace the mass distribution of early-type galaxies.

The first use of PN as a tracer of the kinematics of galaxies was by Meatheringham *et al.* (1988) who used the measured velocities of 94 PN in the LMC to obtain the rotation curve of that galaxy. The HI rotation curve of the LMC can be fitted by a solid-body rotation model out to 2° from the center with an exponential disk outside, and Meatheringham *et al.* (1988) found no significant difference between the HI and PN kinematics other than an increase in velocity dispersion in the older PN population.

20.4.1 Halo dynamics of early type galaxies

This technique of PN kinematics was applied to the giant elliptical galaxy NGC 5128 (Cen A) by Hui *et al.* (1995). The radial velocities of 433 PN were measured, extending as far as 25 kpc from the nucleus. In comparison, stellar kinematics can only reach out to ~5 kpc. NGC 1399, a cD giant elliptical in the Fornax cluster, was observed by Arnaboldi *et al.* (1994), and the velocities of 37 PN in the outer region were used to determined the halo dynamics of this galaxy. Arnaboldi *et al.* (1996) extended this technique to NGC 4406, a non-cD giant in the Virgo cluster. The velocities of 16 PN were measured.

Given an adequate sample, the rotation axis on the sky can be obtained by examining the velocity variation as a function of azimuthal angle. Figure 20.8 shows the PN velocity field in Cen A. It is obvious from this diagram that the axis of zero rotation (the rotation axis) lies somewhere between the photometric major and minor axes. When the velocity data are fitted with a simple sinusoidal function, the rotation axis (shown in Fig. 20.8 as an arrow) is 39° east of the minor axis. This large difference between the photometric and dynamical axes shows that Cen A is not an axisymmetric system, but instead a triaxial system.

The major axis rotation curve can be constructed by folding the PN over an axis perpendicular to the rotation axis (NE onto the SW in Fig. 20.8) and binned into appropriate intervals. The mean velocity and its standard deviation then give the rotation velocity $[V_{\text{rot}}(r)]$ and velocity dispersion $[\sigma(r)]$ for each bin. Figure 20.9 shows that along the major axis, the rotation increases with radius in the inner galaxy and reaches a plateau of ~ 100 km s^{-1} at >7 kpc.

20.4.2 Mass distribution of the galaxy

Assuming that the velocity dispersion is isotropic, then the velocity dispersion and rotational velocity along the major axis obey the spherical Jeans equation:

$$\frac{d[\rho(r)\sigma^2(r)]}{dr} = -\frac{GM(r)\rho(r)}{r^2} + \frac{\rho(r)V_{\text{rot}}^2(r)}{r} \tag{20.3}$$

where $\rho(r)$ is the mass density of test particles and $M(r)$ is the total mass within radius r. The rotation curve can be assumed to have the form

$$V_{\text{rot}}(r) = \frac{v_0 r}{\sqrt{r^2 + r_0^2}}, \tag{20.4}$$

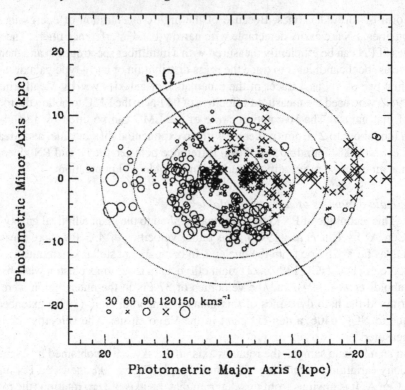

Fig. 20.8. The PN spatial distribution in Cen A. Circles and crosses represent approaching and receding PNm respectively. The ellipses are the appropriate optical isophotes of the galaxy, and the arrow marks the position of the rotation axis (from Hui *et al.*, 1995).

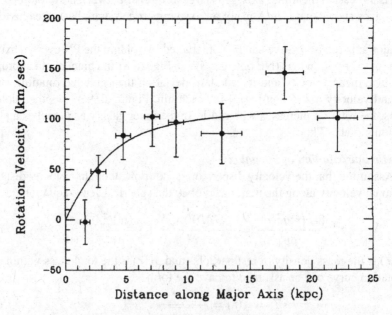

Fig. 20.9. The PN major axis rotation curve for Cen A (from Hui *et al.*, 1995).

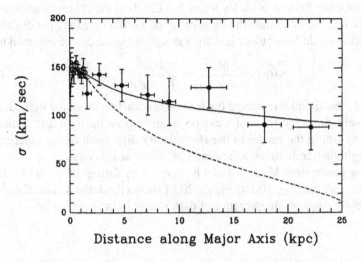

Fig. 20.10. The PN velocity dispersion along the major axis of Cen A. The solid curve is the two-component model given by Eq. (20.5).

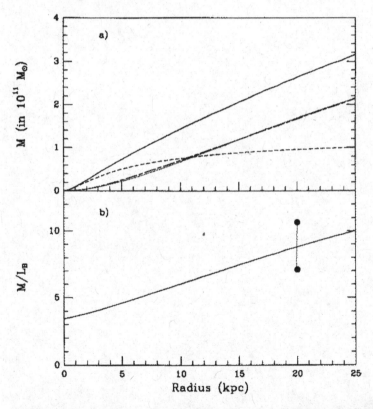

Fig. 20.11. The distribution of mass in Cen A. The upper panel shows the distribution of total mass (solid curve), the luminous component (dash curve), and the dark component (dotted curve) as a function of radius. The lower panel shows the change in the mass to light ratio of the galaxy. The two dots linked by a dotted line are the M/L_B derived using globular cluster data (from Hui *et al.*, 1995).

which is asymptotically flat at $r > r_0$. By fitting Eq. (20.4) to the PN major axis rotation curve (Fig. 20.9), one can determine the value of r_0 and v_0. Because the light distribution of the galaxy follows a de Vaucouleur law, the mass distribution can be assumed to be

$$M(r) = \frac{M_l r^2}{(r+a)^2} + \frac{M_d r^2}{(r+d)^2} \qquad (20.5)$$

where M_l and M_d and a and d represent the mass and scale lengths of the light and dark matter, respectively. Since the value of a can be obtained from the light distribution, M_l can be derived by fitting the model to the Hα velocity dispersion of the inner region, assuming that light matter dominates the center of elliptical galaxies.

The remaining parameters M_d and d can be derived by fitting the model to the PN velocity dispersion profile (Fig. 20.10). Figure 20.11 shows the derived mass distribution for Cen A. It clearly suggests the presence of dark matter in the halo.

21

Concluding remarks

In this book, we have addressed three aspects of PN research: radiation mechanisms, evolution, and applications. We have discussed the various physical mechanisms that are responsible for the emission of radiation from PN, and how different techniques (imaging, photometry and spectroscopy) in the radio, submillmeter, infrared, optical, ultraviolet, and X-ray can be used to probe the physical conditions in different parts of the nebulae. The wealth of data obtained through multi-wavelength observations have served as laboratories for the testing of radiation theories as well as for atomic and molecular physics. The discoveries of new phenomena such as forbidden lines and unidentified infrared features have stimulated the laboratory spectroscopy of atomic and molecular species.

The central stars of PN are hot and luminous objects. Although most of the starlight is in the form of high-energy UV photons, the nebulae are able to intercept these photons and downgrade them to visible wavelengths. Through photoionization, the radiative energy of the star is transferred to the kinetic energy pool of the gaseous nebula, which then emits low-energy line and continuum photons through the processes of recombination and collisional excitation. As the result, PN become bright visible objects and can be detected at large (cosmological) distances. For nearby PN, it is possible to obtain very high-quality spectra which allow for the accurate determination of chemical abundances and kinematic structure of the nebulae.

We have also discussed the evolution of PN using stellar evolution and dynamical models. The evolution of PN is traced from their progenitors, the AGB stars and PPN, to their descendants, the WDs. Part of our modern understanding of PN is the result of our realization that in addition to radiative interactions, there are also mechanical interactions between the star (through its stellar wind) with the nebulae. The physics of stellar structure and nuclear processes also come into play because the star is not static but undergoes significant evolution in temperature and luminosity through the (short) lifetime of PN. The time-dependent radiative and mechanical outputs from the central star therefore lead to changing interactions with the nebulae, resulting in changing morphologies of the nebulae caused by ionization and dynamical effects.

In spite of their short life span, PN play an important role in the determination of mass return to the Galaxy and the chemical enrichment of the Galaxy. Although nucleosynthesis takes place in the AGB phase, the ionization and excitation of atoms in PN make many elements easier to detect and their abundances possible to measure. PN are not only useful in testing the theories of nucleosynthesis; they can also be used to trace the original chemical environment from which they were formed. As such, they represent powerful tools to study the evolution of galaxies.

Research on PN, a subject with over one hundred years of history, has undergone a renaissance in recent years. New observations from space (*IUE, IRAS, ROSAT, HST, ISO,* etc.) have revealed a more complicated and interesting system than previously believed. PN are active in all spectral regions from the X-ray to the radio, radiating by a rich variety of physical mechanisms. They represent the best-studied example of the interactions (radiative, mechanical, and chemical) between stars and the interstellar medium. Lessons that we learned in PN can also be extended to the study of other astrophysical objects. Photoionization models developed for PN have been applied to the study of active galactic nuclei. PN and young stellar objects share many similar characteristics: infrared excess, circumstellar disks, bipolar morphologies, and even microstructures such as knots. The study of PN has been extended from the Milkyway to the Magellanic Clouds and M31 and, more recently, to external galaxies beyond the local group. With the new 8-m class telescopes now coming on line, PN can be found in large numbers against the background stars thanks to their strong monochromatic emission. PN have been successfully used as standard candles to measure the size of the universe, and provide perhaps the best tracers we have of dark matter in galactic halos, or even of the intra-cluster medium. Spectrophotometric studies reveal the abundances of lighter elements such as N, O, Ne and S in the intermediate and old stellar populations. With these recent advances, we expect that PN research will continue to play a major role in modern astrophysics well into the 21st century.

References

Aaquist, O. B., & Kwok, S. 1990, *A&AS*, **84**, 229

Aaquist, O. B., & Kwok, S. 1991, *ApJ*, **378**, 599

Aaquist, O. B., & Kwok, S. 1996, *ApJ*, **462**, 813

Abell, G. O. 1966, *ApJ*, **144**, 259

Abell, G. O., & Goldreich, P. 1966, *PASP*, **78**, 232

Abbott, D. C. 1978, *ApJ*, **225**, 893

Acker, A. 1978, *A&AS*, **33**, 367

Acker, A. 1997, *IAU Symp 180: Planetary Nebulae*, eds. H. J. Habing & H. Lamers (Kluwer:Dordrecht), p. 10

Acker, A., Chopinet, M., Pottasch, S. R., & Stenholm, B. 1987, *A&AS*, **71**, 163

Acker, A., Köppen, J., Stenholm, B., & Jasniewicz, G. 1989, *A&AS*, **80**, 201

Acker, A., Stenholm, B., Tylenda, R., & Raytchev, B. 1991, *A&AS*, **90**, 89

Acker, A., Ochsenbein, F., Stenholm, B., Tylenda, R., Marcout, J., Schohn, C. 1992, *Strasbourg-ESO catalog of Galactic Planetary Nebulae*, ESO publications

Acker, A., Marcout, J., Ochsenbein, F., *et al.* 1996, *First Supplement to the SECGPN*, Observatoire de Strasbourg

Aitken, D. K., & Roche, P. F. 1982, *MNRAS*, **200**, 217. *ApJ*, **233**, 925

Alcolea, J., & Bujarrabal, V. 1991, *A&A*, **245**, 499

Allamandola, L. J., Tielens, A. G. G. M., & Barker, J. R. 1989, *ApJS*, **71**, 733

Aller, L. H. 1956, *Gaseous Nebulae*, (New York: Wiley)

Aller, L. H., & Liller, W. 1968, in *Nebulae and Interstellar Matter*, eds. B. M. Middlehurst, & L. H. Aller (U. of Chicago Press), p. 483

Aller, L. H., & Czyzak, S. J. 1983, *ApJS*, **51**, 211

Aller, L. H., & Keyes, C. D. 1987, *ApJS*, **65**, 405

Aller, L. H., & Milne, D. K. 1972, *Australian J. Physics*, **25**, 91

Aller, L. H., Ufford, W., & Van Vleck, J. H. 1949, *ApJ*, **109**, 42

Aller, L. H., Bowen, I. S., & Minkowski, R. 1955, *ApJ*, **122**, 62

Aller, L. H., Bowen, I. S., & Wilson, O. C. 1963, *ApJ*, **138**, 1013

Ambartsumyan, 1932, *Pulkovo Obs. Circ.*, Vol. **8**, No. 4

Amnuel, P. R., Guseinov, O. Kh., Novruzova, Kh. I, & Rustamov, Iu.S. 1984, *Ap&SS*, 107, 19

Apparao, K. M. V., & Tarafdar, S. P. 1989, *ApJ*, **344**, 826

Arnaboldi, M., *et al.*, 1996, *ApJ*, **472**, 145

Arnaboldi, M., Freeman, K. C., Hui, X., Capaccioli, M., & Ford, H. 1994, *ESO Messenger*, **76**, 40

Baade, W. 1955, *AJ*, **60**, 151

Bachiller, R., Huggins, P. J., Cox, P., & Forveille, T. 1994, *A&A*, **281**, L93

Bachiller, R., Forveille, T., Huggins, P. J., & Cox, P. 1997, *A&A*, **324**, 1123

Bachiller, R., Gomez-Gonzalez, J., Bujarrabal, V., Martin-Pintado, J. 1988, *A&A*, **196**, L5

Bachiller, R., Planesas, P., Martin-Pintado, J., Bujarrabal, V., & Tafalla, M. 1989a, *A&A*, **210**, 366

Bachiller, R., Bujarrabal, V., Martin-Pintado, J., & Gómez-González, J. 1989b, *A&A*, **218**, 252

Bakker, E. J., van Dishoeck, E. F., Waters, L. B. F. M., & Schoenmaker, T. 1997 *A&A*, **323**, 469

Balick B. 1987, *AJ* **94**, 671

Balick, B., Perinotto, M., Macchioni, A., Terzian, Y., & Hajian, A. 1994, *ApJ*, **424**, 800

Balick, B., Alexander, J., Hajian, A., Terzian, Y., Perinotto, M., & Patriarchi, P. 1997, in *IAU Symp. 180: Planetary Nebulae*, eds. H. J. Habing & H. Lamers (Kluwer:Dordrecht), p. 138

Barker, J. R., Allamandola, L. J., & Tielens, A. G. G. M. 1987, *ApJ*, **315**, L61

Barlow, M. J. 1987, *MNRAS*, **227**, 161

Bässgen, M., & Grewing, M. 1989, *A&A*, **218**, 273

Baud, B., & Habing, H. J. 1983, *A&A*, **127**, 73

Bedogni, R., & d'Ercole, A. 1986, *A&A*, **157**, 101

Begemann, B., Dorschner, J., Henning, Th., & Mutschke, H. 1996, *ApJ*, **464**, L195

Becker, S. A., & Iben, I. Jr. 1979, *ApJ*, **232**, 831

Becker, S. A., & Iben, I. Jr. 1980, *ApJ*, **237**, 111

Beckwith, S., Persson, S. E., & Gatley, I. 1978, *ApJ*, **219**, L33

Bergeron, P., Saffer, R., & Liebert, J. 1992, *ApJ*, **394**, 228

Bianchi, L., Cerrato, S., & Grewing, M. 1986, *A&A*, **169**, 227

Bidelman, W. P. 1985, *BAAS*, **17**, 842

Bidelman, W. P. 1986, in *IAU Symp. 122: Circumstellar Matter*, eds. I. Appenzeller, & C. Jordan (Reidel:Dordrecht), p. 217

Bieging, J. H., Wilner, D., & Thronson, H. A. 1991, *ApJ*, **379**, 271

Black, J., & van Dishoeck, E. F. 1987, *ApJ*, **322**, 412

Blöcker, T. 1995a, *A&A*, **297**, 727

Blöcker, T. 1995b, *A&A*, **299**, 755

Blöcker, T., & Schönberner, D. 1990, *A&A*, **240**, L11

Blöcker, T., & Schönberner, D. 1991, *A&A*, **244**, L23

Blöcker, T., Herwig, F., Driebe, T., Bramkamp, H., & Schönberner, D. 1997, in *IAU Symp. 180: Planetary Nebulae*, eds. H. J. Habing, & H Lamers (Kluwer:Dordrecht), p. 389

Blondin, J. M., & Lundqvist, P. 1993, *ApJ*, **405**, 337

Bohm, D., & Aller, L. H. 1947, *ApJ*, **105**, 131

Bond, H. E. 1995, in *Asymmetric Planetary Nebulae, Ann. Israel Phys. Soc.* **Vol. 11**, eds. A. Harpaz, & N. Soker, p. 61

Bond, H. E. 1997, in *IAU Symp. 180: Planetary Nebulae*, eds. H. J. Habing & H. Lamers (Kluwer:Dordrecht), p. 460

Bond, H. E., Luck, R. E. 1987. In *IAU Colloq. 95: The Second Conference on Faint Blue Stars*, eds. A. G. D. Philips, D. S. Hayes & J. W. Liebert, p. 527

Bond, H. E., & Meakes, M. G. 1990, *AJ*, **100**, 788

Bond, H. E., *et al.*, 1996, *AJ*, **112**, 2699

Boothroyd, A. I., & Sackmann, I.-J. 1988a, *ApJ*, **328**, 632

Boothroyd, A. I., & Sackmann, I.-J. 1988b, *ApJ*, **328**, 641

Boothroyd, A. I., & Sackmann, I.-J. 1988c, *ApJ*, **328**, 653

Boothroyd, A. I., & Sackmann, I.-J. 1988d, *ApJ*, **328**, 671

Boothroyd, A. I., & Sackmann, I.-J. 1992, *ApJ*, **393**, L21

Borkowski, K. J., Harrington, J. P., & Tsvetanov, Z. I. 1995, *ApJ*, **449**, L143

Bowen, I. S. 1928, *ApJ*, **67**, 1

Bowen, I. S. 1935, *ApJ*, **81**, 1

Bowen, G. H. 1988, *ApJ*, **329**, 299

Bowers, P. F. 1985, in *Mass Loss from Red Giants*, eds. M. Morris, & B. Zuckerman (Reidel:Dordrecht), p. 189

Breitschwerdt, D., & Kahn, F. D. 1990, *MNRAS*, **244**, 521

Brocklehurst, M. 1970, *MNRAS*, **148**, 417

Brocklehurst, M. 1971, *MNRAS*, **153**, 471

Brown, R. L., & Mathews, W. G. 1970, *ApJ*, **160**, 939

Bryan, G. L., Volk, K., & Kwok, S. 1990, *ApJ*, **365**, 301

Buckley, D., & Schneider, S. E. 1995, *ApJ*, **446**, 279

Burgess, A. 1965, *Mem RAS*, **69**, 1

Bujarrabal, V., Alcolea, J., & Planesas, P. 1992, *A&A*, **257**, 701

Bujarrabal, V., Bachiller, R., Alcolea, J., & Martin-Pintado, J. 1988, *A&A*, **206**, L17

Buss, R. H., Cohen, M., Tielens, A. G. G. M., Werner, M. W., Bregman, J. D., Rank, D., Witteborn, F. C., & Sandford, S. 1990, *ApJ*, **365**, L23

Cahn, J. H., & Kaler, J. B. 1971, *ApJS*, **22**, 319

Cahn, J. H., & Wyatt, S. P. 1976, *ApJ*, 210, 508

Cahn, J. H., Kaler, J. B., & Stanghellini, L. 1992, *A&AS*, **94**, 399

Cai, W., & Pradhan, A. K. 1993, *ApJS*, **88**, 329

Capriotti, E. R. 1973, *ApJ*, **179**, 495

Capriotti, E. R. 1978, in *IAU Symp. 76: Planetary Nebulae*, ed. Y. Terzian (Reidel:Dordrecht), p. 263

Carranza, G., Courtès, G., & Louise, R. 1968, in *IAU Symp. 34: Planetary Nebulae*, eds. D. Osterbrock, & C. R. O'Dell, (Reidel:Dordrecht), p. 249

Carsenty, U., & Solf, J. 1982, *A&A*, **106**, 307

Cassinelli, J. 1979, *Ann. Rev. Astr. Ap.*, **17**, 275

Castor, J. I. 1970, *MNRAS*, **149**, 111

Castor, J. 1981, in *Physical Processes in Red Giants*, eds. I. Iben & A. Renzini (Reidel:Dordrecht), p. 285

Castor, J. I., & Lamers, J. G. L. M. 1979, *ApJS*, **39**, 481

Castor, J. I., Lutz, J. H., & Seaton, M. J. 1981, *MNRAS*, **194**, 547

Cernoicharo, J., Guélin, M., Martin-Pintado, J., Penalver, J., & Mauersberger, R. 1989, *A&A*, **222**, L1

Cerruti-Sola, M., & Perinotto, M. 1985, *ApJ*, **291**, 237

Chu, Y. 1989, *IAU Symp 131: Planetary Nebulae*, ed. S. Torres-Peimbert (Reidel:Dordrecht), p. 105

Chu, Y.-H. 1993, in *IAU Symp. 155: Planetary Nebulae*, eds. R. Weinberger & A. Acker (Kluwer:Dordrecht), p. 139

Chu, Y., Jacoby, G. H., & Arendt, R. 1987, *ApJS*, **64**, 529

Chu, Y.-H., Kwitter, K. B., & Kaler, J. B. 1993, *AJ*, **106**, 650

Chu, Y., Kwitter, K. B., Kaler, J. B., & Jacoby, G. H. 1984, *PASP*, **96**, 598

Chu, Y.-H., Chang, T. H., & Conway, G. M. 1997, *ApJ*, **482**, 891

Ciardullo, R., & Bond, H. E. 1996, *AJ*, **111**, 2332

Ciardullo, R., Jacoby, G. H., Ford, H. C., Neill, J. D. 1989, *ApJ*, **339**, 53

Clegg, R. E. S. 1987, *MNRAS*, **221**, 31P

Clegg, R. E. S. 1991, in *IAU Symp. 145, Evolution of Stars: The Photospheric Abundance Connection*, eds. G. Michard, and A. Tutukov (Kluwer:Dordrecht), p. 387

Clegg, R. E. S. 1997, in *IAU Symp. 180: Planetary Nebulae*, eds. H. J. Habing & H. Lamers (Kluwer:Dordrecht), p. 549

Clegg, R. E. S., & Middlemass, D. 1987, *MNRAS*, **228**, 759

Clegg, R. E. S., Storey, P. J., Walsh, J. R., & Neale, L. 1997, *MNRAS*, **284**, 348

Cohen, M., & Barlow, M. J. 1974, *ApJ*, **193**, 401

Cohen, M., & Barlow, M. J. 1975, *ApL*, 16, 165

Cohen, M., & Barlow, M. J. 1980, *ApJ*, **238**, 585

Cohen, M., & Kuhi, L. V. 1980, *PASP*, **92**, 736

Cohen, M. *et al.*, 1975, *ApJ*, **196**, 179

Cohen, M. *et al.*, 1989, *ApJ*, **341**, 246

Conway, G. M., & Chu, Y.-H. 1997, in *IAU Symp. 180: Planetary Nebulae*, eds. H. J. Habing, & H. Lamers (Kluwer:Dordrecht), p. 214

Corradi, R. L. M., & Schwarz, H. E. 1995, *A&A*, **293**, 871

Cox, P. 1991, *A&A*, **236**, L29

Cox, P. 1993, in *Astronomical Infrared Spectroscopy*, ed. S. Kwok, ASP Conf. Ser. **41**, p. 163

Cox, P., Maillard, J.-P., Huggins, P. J., Forveille, T., Simons, D., Guilloteau, S. Rigaut, F., Bachiller, R., & Omont, A., 1997, *A&A*, **321**, 907

Crampton, D., Cowley, A. P., & Humphreys, R. M. 1975. *ApJ*, **198**, L135

Curtis, H. D. 1918, *Publ. Lick Obs.*, Vol. XIII, Part III, p. 57

Czyzak, S. J., & Aller, L. H. 1970, *ApJ*, **162**, 495

Daub, C. T. 1982, *ApJ*, **260**, 612

Davis, L. E., Seaquist, E. R., & Purton, C. R. 1979, *ApJ*, **230**, 434

Decin, L., Van Winckel, H., Waelkens, C., & Bakker, E. J. 1998, *A&A*, **332**, 928

Deguchi, S. 1995, in *Asymmetric Planetary Nebulae, Ann. Israel Phys. Soc.* **Vol. 11**, eds. A. Harpaz, & N. Soker, p. 132

Deinzer, W. 1967, *Zeitschr. Astrophys.*, **67**, 342

Deinzer, W., & von Sengbusch, K. 1970, *ApJ*, **160**, 671

Deutsch, A. J. 1956, *ApJ*, **123**, 210

Dinerstein, H. L., & Sneden, C. 1988, *ApJ*, **335**, L23

Dinerstein, H. L., Sneden, C., McQuitty, R., Danly, L., & Heap, S. 1994, *BAAS*, **26**, 1386

Dinerstein, H. L., Lester, D. F., Carr, J. S., & Harvey, P. M. 1988, *ApJ*, **327**, L27

Dopita, M. A., & Hua, C. T. 1997, *ApJS*, **108**, 515

Dopita, M. A., & Meatheringham, S. J. 1990, *ApJ*, **357**, 140

Dopita, M. A., & Meatheringham, S. J. 1991a, *ApJ*, **367**, 115

Dopita, M. A., & Meatheringham, S. J. 1991b, *ApJ*, **377**, 480

Dopita, M. A., Meatheringham, S. J., Wood, P. R., Webster, B. L., Morgan, D. H., & Ford, H. C. 1987, *ApJ*, **315**, L107

Dopita, M. A., Meatheringham, S. J., Webster, B. L., & Ford, H. C. 1988, *ApJ*, **327**, 639

Dopita, M. A., Vassiliadis, E., Meatheringham, S. J., Bohlin, R. C., Ford, H. C., Harrington, J. P., Wood, P. R., Stecher, T. P., & Maran, S. P. 1996, *ApJ*, **460**, 320

Dorman, B., Rood, R. T., & O'Connell, R. W. 1993, *ApJ*, **419**, 596

Drilling, J. S., & Schönberner, D. 1985, *A&A*, **146**, L23

Duley, W., & Williams, D. A. 1986, *MNRAS*, **219**, 859

Duncan, J. C. 1937, *ApJ*, **86**, 496

Dyson, J. E. 1981, in *Investigating the Universe*, ed. F. D. Kahn, (Reidel:Dordrecht), p. 125

Dyson, J. E. 1984 *Ap&SS* **106**, 181

Eder, J., Lewis, B. M., & Terzian, Y. 1988, *ApJS*, **66**, 183

Elson, R. A. W., Sigurdsson, S., Hurley, J., Davies, M. B., & Gilmore, G. F. 1998, *ApJ*, **499**, L53

Esteban, C., Smith, L. J., Vìchez, J. M., & Clegg, R. E. S. 1993, *A&A*, **272**, 299

Eulderink, F., & Mellema, G. 1994, *A&A*, **284**, 654

Faulkner, D. J. 1970, *ApJ*, **162**, 513

Faúndez-Abans, M., & Maciel, W. J. 1986, *A&A*, **158**, 228

Feast, M. 1987, *The galaxy*, NATO ASI Series, eds. G. Gilmore, & B. Carswell, (Reidel:Dordrecht), p. 1

Feast, M., Catchpole, R. M., Whitelock, P. A., Carter, B. S., & Roberts, G. 1983, *MNRAS*, **203**, 373

Feldmeier, J. J., Ciardullo, R., & Jacoby, G. 1996, *ApJ*, **461**, L25

Ferch, R. L. & Salpeter, E. E. 1975, *ApJ*, **202**, 195

Ferguson, H. C., & Davidsen, A. F. 1993, *ApJ*, **408**, 92

Ferland, G. J. 1993, in *IAU Symp. 155: Planetary Nebulae*, eds. R. Weinberger & A. Acker (Kluwer:Dordrecht), p. 123

Feibelman, W. A. 1994, *PASP*, **106**, 756

Feibelman, W. A. 1999, *ApJ*, **514**, 296

Finzi, A., & Wolf, R. A. 1971, *A&A*, **11**, 418

Ford, H. C., & Jacoby, G. H. 1978, *ApJ*, **219**, 437

Ford, H. C., & Jenner, D. C. 1978, *BAAS*, **10**, 665

Forveille, T., Morris, M., Omont, A., & Likkel, L. 1987, *A&A*, **176**, L13

Frank, A. 1994, *AJ*, **107**, 261

Frank, A., & Mellema, G. 1994a, *ApJ*, **430**, 800

Frank, A., & Mellema, G. 1994b, *A&A*, **289**, 937

Frank, A., & Noriega-Crespo, A. 1994, *A&A*, **290**, 643

Frank, A., van der Veen, W. E. C. J., & Balick, B. 1994, *A&A*, **282**, 554

Frank, A., Balick, B., Icke, V., & Mellema, G. 1993, *ApJ*, **404**, L25

Fruscione, A., Drake, J. J., McDonald, K., & Malina, R. F. 1995, *ApJ*, **441**, 726

Gammie, C. F., Knapp, G. R., Young, K., Phillips, T. G., & Falgarone, E. 1989, *ApJ*, **345**, L87

Gathier, R. 1984, PhD thesis, University of Groningen

Gathier, R. 1987, *A&AS*, **71**, 245

Gathier, R., & Pottasch, S. R. 1989, *A&A*, **209**, 369

Gathier, R., Pottasch, S. R., & Pel, J. W. 1986, *A&A*, **157**, 171

Gathier, R., Pottasch, S. R., & Goss, W. M. 1986, *A&A*, **157**, 191

Gathier, R., Pottasch, S. R., Goss, W. M., & Van Gorkom, J. H. 1983, *A&A*, **128**, 325

Gatley, I., DePoy, D. L., & Flower, A. M. 1988, *Science*, **242**, 1264

Geballe, T. R., & van der Veen, W. E. C. J. 1990, *A&A*, **235**, L9

Geballe, T. R., Kim, Y. H., Knacke, R. F., & Noll, K. S. 1988, *ApJ*, **326**, L65

Geballe, T. R., Tielens, A. G. G. M., Kwok, S., & Hrivnak, B. J. 1992, *ApJ*, **387**, L89

Geballe, T. R., Lacy, J. H., Persson, S. E., McGregor, P. J., & Soifer, B. T. 1985, *ApJ*, **292**, 500

Geballe, T. R., Joblin, C., d'Hendecourt, L. B., Jourdain de Muizon, M., Tielens, A. G. G. M., & Leger, A. 1994, *ApJ*, **434**, L15

Gehrz, R. D. 1972, *ApJ*, **178**, 715

Gehrz, R. D., & Woolf, N. J. 1970, *ApJ*, **161**, L213

Gehrz, R. D., & Woolf, N. J. 1971, *ApJ*, **165**, 285

George, D., Kaftan-Kassim, M. A., & Hartsuijker, A. P. 1974, *A&A*, **35**, 219

Gillett, F. C., Low, F. J., & Stein, W. A. 1967, *ApJ*, **149**, L97

Gillett, F. C., Hyland, A. R., & Stein, W. A. 1970, *ApJ*, **162**, L21

Gillett, F. C., Backman, D. E., Beichman, C., & Neugebauer, G. 1986, *ApJ*, **310**, 842

Gilman, R. C. 1969, *ApJ*, **155**, L185

Gilman, R. C. 1972, *ApJ*, **178**, 423

Gleizes, F., Acker, A., & Stenholm, B. 1989, *A&A*, **222**, 237

Goebel, J. H. 1993 *A&A*, **278**, 226

Górny, S. K., & Stasińska, G. 1995, *A&A*, **303**, 893

Gottlieb, E. W., & Liller, W. 1976, *ApJ*, **207**, L135

Graham, J. F., Serabyn, E., Herbst, T. M., Matthews, K., Neugebauer, G., Soifer, B. T., Wilson, T. D., & Beckwith, S. 1993, *AJ*, **105**, 250

Grauer, A. D., & Bond, H. E. 1984, *ApJ*, **277**, 211

Green, R., Schmidt, M., & Liebert, J. 1986, *ApJS*, **61**, 305

Greenhouse, M. A., Hayward, T. L., & Thronson, H. A. 1988, *ApJ*, **325**, 604

Greggio, L., & Renzini, A. 1990, *ApJ*, **364**, 35

Greig, W. E. 1971, *A&A*, **10**, 161

Hajian, A. R., Terzian, Y. & Bignell, R. C. 1993, *AJ*, **106**, 1965

Hajian, A. R., Frank, A., Balick, B., & Terzian, Y. 1997, *ApJ*, **477**, 226

Hamann, W. R. (1996), *Ap&SS*, **238**, 31

Härm, R., & Schwarzschild, M. 1975, *ApJ*, **200**, 324

Harman, R. J., & Seaton, M. J. 1964, *ApJ*, **140**, 827

Harman, R. J., & Seaton, M. J. 1966, *MNRAS*, **132**, 15

Harrington, J. P., Lutz, J. H., & Seaton, M. J. 1981, *MNRAS*, **195**, 21P

Harrington, J. P., Seaton, M. J., Adams, S., & Lutz, J. H. 1982, *MNRAS*, **199**, 517

Harris, H. C., Dahn, C. C., Monet, D. G., & Pier, J. R. 1997, in *IAU Symp. 180: Planetary Nebulae*, eds. H. J. Habing, & H. Lamers (Kluwer:Dordrecht), p. 40

Hawkins, G. W., Skinner, C. J., Meixner, M. M., Jernigan, J. G., Arens, J. F., Keto, E., & Graham, J. R. 1995, *ApJ*, **452**, 314

Heap, S. R. 1982, in *IAU Symp 99: Wolf-Rayet Stars, observations, physics, evolution.* eds. C. W. H. de Loore, & A. J. Willis (Reidel:Dordrecht), p. 423

Heap, S. R. 1993, in *IAU Symp. 155: Planetary Nebulae*, eds. A. Acker & R. Weinberger (Kluwer:Dordrecht), p. 23

Heap, S. R., & Augensen, H. J. 1987, *ApJ*, **313**, 268

Heap, S. R., & Hintzen, P. 1990, *ApJ*, **353**, 200

Heap, S. R., *et al.*, 1978, *Nature*, **275**, 385

Heber, U., Dreizler, S., & Hagen, H.-J. 1996, *A&A*, **311**, L17

Heiligman, G. M. 1986, *ApJ*, **308**, 306

Henize, K. G. 1967, *ApJS*, **14**, 125

Henry, R. B. C. 1989, *MNRAS*, **241**, 453

Henry, R. B. C., & Shipman, H. L. 1986, *ApJ*, **311**, 774

Herman, J., & Habing, H. J. 1985a, *Phys. Rep.*, **124**, 255

Herman, J., & Habing, H. J. 1985b, *A&AS*, **59**, 523

Herschel, W. 1791, *Abridged Phil. Trans.*, **17**, 25

Higgs, L. A. 1971, *Publ. Astrophys. Bra. NRC Canada*, **1**, 1

Hill, H. G. M., Jones, A. P., & d'Hendecourt, L. B. 1998, *A&A*, **336**, L41

Hollenbach, D. J., & Tielens, A. G. G. M. 1997, *Ann. Rev. Astr. Ap.*, **35**, 179

Hodge, P. W. 1966, *Galaxies and Cosmology* (McGraw-Hill:New York)

Holzer, T. E., & MacGregor, K. B. 1985, in *Mass Loss from Red Giants*, eds. M. Morris & B. Zuckerman (Reidel:Dordrecht), p. 229

Howe, D. A., Millar, T. J., & Williams, D. A. 1992, *MNRAS*, **255**, 217

Hrivnak B. J. 1995, *ApJ* **438**, 341

Hrivnak, B. J. 1997, in *IAU Symp. 180: Planetary Nebulae*, eds. H. J. Habing & H. Lamers (Kluwer:Dordrecht), p. 303

Hrivnak, B. J., & Kwok, S. 1991, *ApJ*, **368**, 564

Hrivnak, B. J., Kwok, S., & Volk, K. 1988, *ApJ*, **331**, 832

Hrivnak, B. J., Kwok, S., Volk, K. 1989, *ApJ*, **346**, 265

Hrivnak, B. J., Kwok, S., & Geballe, T. R. 1994, *ApJ*, **420**, 783

Hua, C. T. 1974, *A&A*, **32**, 423

Hua, C. T. 1997, *A&AS*, **125**, 355

Hua, C. T., & Louise, R. 1970, *A&A*, **9**, 448

Hua, C. T., & Grundseth, B. 1986, *AJ*, **92**, 853

Hua, C. T., Dopita, M. A., & Martinis, J. 1998, *A&AS*, **133**, 361

Hubble, E. 1922, *ApJ*, **56**, 400

Hui, X., Ford, H. C., Freeman, K. C., & Dopita, M. A. 1995, *ApJ*, **449**, 592

Huggins, P. J., & Healy, A. P. 1989, *ApJ*, **346**, 201

Hummer, D. G., & Storey, P. J. 1987, *MNRAS*, **224**, 801

Husfield, D., Kudritzki, R. P., Simon, K. P., & Clegg, R. E. S. 1984, *A&A*, **134**, 139

Hyung, S., & Aller, L. H. 1997, *ApJ*, **491**, 242

Hyung, S., Aller, L. H., & Feibelman, W. A. 1994a, *ApJS*, **93**, 465

Hyung, S., Aller, L. H., & Feibelman, W. A. 1994, *MNRAS*, **269**, 975

Iben, I. 1984, *ApJ*, **277**, 333

Iben, I. 1991, *ApJS*, **76**, 55

Iben, I. 1995, *Phys. Reports*, **250**, 1

Iben, I., & Livio, M. 1993, *PASP*, **105**, 1373

Iben, I., & Renzini, A. 1983, *Ann. Rev. A&Ap*, **21**, 271

Iben, I., & Renzini, A. 1984, *Phys. Rep.* **105**, 329

Icke, V. 1988, *A&A*, **202**, 177

Ishida, K., & Weinberger, R. 1987, *A&A*, **178**, 227

Jacoby, G. 1989, *ApJ*, **339**, 39

Jacoby, G. H. 1997, in *IAU Symp. 180: Planetary Nebulae*, eds. H. J. Habing & H. Lamers (Kluwer:Dordrecht), p. 448

Jacoby, G., Morse, J. A., Fullton, L. K., Kwitter, K. B., & Henry, R. B. 1997, *AJ*, **114**, 2611

Jacoby, G., *et al.* 1992, *PASP*, **104**, 599

Jacoby, G., Ciardullo, R., & Harris, W. E. 1996, *ApJ*, **462**, 1

Jaminet, P. A., Danchi, W. C., Sandell, G., & Sutton, E. C. 1992, *ApJ*, **400**, 535

Jewitt, D. C., Danielson, G. E., & Kupferman, P. N. 1986, *ApJ*, **302**, 727

Jourdain de Muizon, M., d'Hendercourt, L. B., & Geballe, T. R. 1989, *Infrared Spectroscopy in Astronomy*, **ESA SP-290**, p. 177

Justtanont, K., Barlow, M. J., Skinner, C. J., Rche, P. F., Aitken, D. K., & Smith, C. H. 1996, *A&A*, **309**, 612

Kahn, F. D. 1976, *A&A*, **50**, 145

Kahn, F. D. 1983, in *IAU Symp. 103: Planetary Nebulae*, ed. D. R. Flower (Reidel:Dordrecht), p. 305

Kahn, F., & West, K. A. 1985, *MNRAS*, **212**, 837

Kahn, F., & Breitschwerdt, D. 1990, *MNRAS*, **242**, 505

Kaler, J. B. 1974, *AJ*, **79**, 594

Kaler, J. B. 1976, *ApJ*, **210**, 843

Kaler, J. B. 1983, *ApJ*, **271**, 188

Kaler, J. B., & Jacoby, G. H. 1989, *ApJ*, **345**, 871

Kaler, J. B., & Jacoby, G. H. 1990, *ApJ*, **362**, 491

Kaler, J. B., & Jacoby, G. H. 1991, *ApJ*, **382**, 134

Kaler, J. B., & Lutz, J. H. 1985, *PASP*, **97**, 700

Kaler, J. B., Shaw, R. A., & Kwitter, K. B. 1990, *ApJ*, **359**, 392

Kaler, J. B., Aller, L. H., Czyzak, S. J., & Epps, H. W. 1976, *ApJS*, **31**, 163

Karzas, W. J., & Latter, R. 1961, *ApJS*, **6**, 167

Kastner, J. H., Weintraub, D. A., Gatley, I., Merrill, K. M., & Probst, R. G. 1996, *ApJ*, **462**, 777

Kawaler, S. D. 1996, in *Stellar Remnants*, (Springer:Berlin), p. 1

Kennicutt, R. C., Freedman, W. L., & Mould, J. R. 1995, *AJ*, **110**, 1476

Khromov, G. S., & Kohoutek, L. 1968, in *IAU Symp. 34: Planetary Nebulae*, eds. D. E. Osterbrock, & C. R. O'Dell (Reidel:Dordrecht), p. 227

Kingsburgh, R. L., & Barlow, M. J. 1992, *MNRAS*, **257**, 317

Kingsburgh, R. L., & Barlow, M. J. 1994, *MNRAS*, **271**, 257

Knapp, G. R. 1987, in *Late Stages of Stellar Evolution*, eds. S. Kwok & S. R. Pottasch (Reidel:Dordrecht), p. 103

Knapp, G. R., & Morris, M. 1985, *ApJ*, **292**, 640

Knapp, G. R., Sandell, G., & Robson, E. I. 1993, *ApJS*, **88**, 173

Knapp, G. R., Phillips, T. G., Leighton, R. B., Lo, K. Y., Wannier, P. G., & Wootten, H. A. 1982, *ApJ*, **252**, 616

Koester, D., & Reimers, D. 1993, *A&A*, **275**, 479

Koester, D., & Reimers, D. 1996, *A&A*, **313**, 810

Kohoutek, L. 1978, *IAU Symp. 76: Planetary Nebulae*, ed. Y. Terzian (Reidel:Dordrecht), p. 47

Kohoutek, L. 1989, *IAU Symp. 131: Planetary Nebulae*, ed. S. Torres-Peimbert (Kluwer:Dordrecht), p. 29

Kohoutek, L. 1994, *Astron. Nachr.*, **315**, 63

Köppen, J. 1979, *A&AS*, **35**, 111

Köppen, J., & Aller, L. H. 1987, in *Scientific Accomplishments of the IUE*, ed. Y. Kondo (Reidel:Dordrecht), p. 589

Köppen, J., Acker, A., & Stenholm, 1991, *A&A*, **248**, 197

Kovetz, A., & Harpaz, A. 1981, *A&A*, **95**, 66

Kreysing, H. C., Diesch, C., Zweigle, J., Staubert, R., & Grewing, M. 1993, *A&A*, **264**, 623

Krishna Swamy, K. S., & O'Dell, C. R. 1968, *ApJ*, **151**, L61

Kudritzki, R. P., Méndez, R. H., Puls, J., & McCarthy, J. K. 1997, in *IAU Symp. 180: Planetary Nebulae*, ed. H. J. Habing, & H. Lamers (Kluwer:Dordrecht), p. 64

Kuiper, T. B. H., Knapp, G. R., Knapp, S. L., & Brown, R. L. 1976, *ApJ*, **204**, 408

Kurucz, R. L. 1991, in *Precision Photometry*, eds. A. G. Davis Philip *et al.*, (L. Davis Press:Schenectady, N.Y.), p. 27

Kutter, G. S., & Sparks, W. M. 1974, *ApJ*, **192**, 447

Kwok, S. 1975, *ApJ*, **198**, 583

Kwok, S. 1976, *J. Roy. Astr. Soc. Canada*, **70**, 49

Kwok, S. 1981a, in *Effects of Mass Loss on Stellar Evolution*, eds. C. Chiosi & R. Stalio (Reidel:Dordrecht), p. 347

Kwok, S. 1981b, in *Physical Processes in Red Giants*, eds. I. Iben & A. Renzini (Reidel:Dordrecht), p. 421

Kwok, S. 1982, *ApJ*, **258**, 280

Kwok, S. 1983, in *IAU Symp 103: Planetary Nebulae*, eds. D. R. Flower (Reidel:Dordrecht), p. 293

Kwok, S. 1985a, *AJ*, **90**, 49

Kwok, S. 1985b, *ApJ*, **290**, 568

Kwok, S. 1988, *in IAU Coll. 103: The Symbiotic Phenomenon*, eds. J. Mikolajewska *et al.* (Kluwer:Dordrecht), p. 129

Kwok, S. 1990, *MNRAS*, **244**, 179

Kwok, S., & Aaquist, O. B. 1993, *PASP*, **105**, 1456

Kwok, S., & Bignell, R. C. 1984. *ApJ*, **276**, 544

Kwok, S., & Feldman, P. A. 1981. *ApJ*, **247**, L67

Kwok, S., & Volk, K. 1985, *ApJ*, **299**, 191

Kwok, S., Purton, C. R., & FitzGerald, M. P. 1978, *ApJ*, **219**, L125

Kwok, S., Hrivnak, B. J., & Milone, E. F. 1986, *ApJ*, **303**, 451

Kwok, S., Hrivnak, B. J., & Boreiko, R. T. 1987, *ApJ*, **321**, 975

Kwok, S., Volk, K., & Hrivnak, B. J. 1989 *ApJ*, **345**, L51

Kwok, S., Su, K. Y. L., & Hrivnak, B. J. 1998, *ApJ*, **501**, L117

Kwok, S., Hrivnak, B. J., & Geballe, T. R. 1995, *ApJ*, **454**, 394

Kwok, S., Hrivnak, B. J., & Langill, P. P. L. 1993, *ApJ*, **408**, 586

Kwok, S., Volk, K., & Bidelman, W. P. 1997, *ApJS*, **112**, 557

Kwok, S., Purton, C. R., Matthews, H. E., & Spoelstra, T. A. Th. 1985, *A&A*, **114**, 321

Lamers, H. J. G. L. M. & Morton, D. C. 1976, *ApJS*, **32**, 715

Lamers, H. J. G. L. M., Waters, L. B. F. M., Garmany, C. D., Perez, M. R., & Waelkens, C. 1986, *A&A*, **154**, L20

Latter, W. B., & Hora, J. L. 1997, in *IAU Symp 180: Planetary Nebulae*, eds. H. J. Habing & H. Lamers (Kluwer:Dordrecht), p. 254

Latter, W. B., Hora, J. L., Kelly, D. M., Deutsch, L. K., Maloney, P. R. 1993, *AJ*, **106**, 260

Latter, W. B., Maloney, P. R., Kelly, D. M., Black, J. H., Rieke, G. H., Rieke, M. J. 1992, *ApJ*, **389**, 347

Latter, W. B., Kelly, D. M., Hora, J. L., & Deutsch, L. K. 1995, *ApJS*, **100**, 159

Leahy, D. A., Zhang, C. Y., & Kwok, S. 1994, *ApJ*, **422**, 205

Leahy, D. A., Zhang, C. Y., Volk, K., & Kwok, S. 1996, *ApJ*, **466**, 352

Leahy, D. A., Kwok, S., & Yin, D. 1998, *BAAS*, **192**, 5317

Lequeux, J., & Jourdain de Muizon, M. 1990, *A&A* **240**, L19

Leung, C. M. 1976, *ApJ*, **209**, 75

Lewis, B. M. 1989, *ApJ*, **338**, 234

Lewis, B. M., Eder, J., & Terzian, Y. 1990, *ApJ*, **362**, 634

Li, P. S., Kwok, S., & Thronson, H. 1999, in preparation

Likkel, L. 1989, *ApJ*, **344**, 350

Likkel, L., Omont, A., Morris, M., & Forveille, T. 1987, *A&A*, **173**, L11

Likkel, L., Morris, M., Kastner, J. H., & Forveille, T. 1994, *A&A*, **282**, 190

Liller, M. H., Welther, B. L., & Liller, W. 1966, *ApJ*, **144**, 280

Livio, M. 1994, in *Circumstellar Media in the Late Stages of Stellar Evolution*, eds. R. E. S. Clegg
    et al. (Cambridge University Press: Cambridge), p. 35

Livio, M. 1995, in *Asymmetric Planetary Nebulae, Ann. Israel Phys. Soc.* **Vol. 11**, eds. A. Harpaz &
    N. Soker, p. 51

Lo, K. Y., & Bechis, K. P. 1976, *ApJ* **205**, L21

López, J. A., Vázquez, R., & Rodriguez, L. F. 1995, *ApJ*, **455**, L63

Loup, C., Forveille, T., Omont, A., & Paul, J. F. 1993, *A&AS*, **99**, 291

Luo, D., & McCray, R. 1991, *ApJ*, **379**, 659

Lucy, L. B. 1967, *AJ*, **72**, 813

Lutz, J. H. 1973, *ApJ*, **181**, 135

Luyten, W. J. (1979) *NLTT catalog (Minneapolis: University of Minnesota)*

Luyten, W., & Herbig, G. H. 1960, *Harvard Announcement Card*, 1474

Maciel, W. J., & Pottasch, S. R. 1980, *A&A*, **88**, 1

MacLow, M. M., McCray, R., & Norman, M. L. 1989, *ApJ*, **337**, 141

Manchado, A., Guerrero, M. A., Stanghellini, L., & Serra-Ricart, M. 1996, *The IAC Morphological Catalog
    of Northern Galactic Planetary Nebulae*, (IAC: Tenerife)

Marten, H., & Schönberner, D. 1991, *A&A*, **248**, 590

Masson, C. R. 1989a, *ApJ*, **336**, 294

Masson C. R. 1989b, *ApJ*, **346**, 243

Masson C. R. 1990, *ApJ*, **348**, 580

Mathews, W. G. 1966, *ApJ*, **143**, 173

McCarthy, J. K., Mould, J. R., Méndez, R. H., Kudritzki, R. P., Husfeld, D., Herrero, A., & Groth, H. G.
    1990, *ApJ*, **351**, 230

Meatheringham, S. J., Dopita, D. A., Ford, H. C., & Webster, B. L. 1988 *ApJ*, **327**, 651

Mellema, G. 1994, *A&A*, **290**, 915

Mellema, G. 1995, *MNRAS*, **277**, 173

Mellema, G. 1997, *A&A*, **321**, L29

Mellema, G., & Frank, A. 1995, *MNRAS*, **273**, 401

Méndez, R. H. 1991, in *IAU Symp 145: Evolution of Stars: the Photospheric Abundance Connection*, eds.
    G. Michaud & A. Tutukov, (Kluwer:Dordrecht), p. 375

Méndez, R. H., Kudritzki, R. P., & Herrero, A. 1992, *A&A*, **260**, 329

Méndez, R. H., Kudrizki, R. P., Herrero, A., Husfeld, D., & Groth, H. G. 1988, *A&AS*, **190**, 113

Menzel, D. H. 1926, *PASP*, **38**, 295

Menzel, D. H., & Aller, L. H. 1941, *ApJ*, **94**, 30

Menzel, D. H., & Baker, J. G. 1937, *ApJ*, **86**, 70

Milikan, A. G. 1974, *AJ*, **79**, 1259

Miller, & Scalo 1979, *ApJS*, **41**, 513

Milne, D. K. 1979, *A&AS*, **36**, 227

Milne, D. K. 1982, *MNRAS*, **200**, 51P

Milne, D. K., & Aller, L. H. 1975, *A&A*, **38**, 183

Milne, D. K., & Aller, L. H. 1982, *A&AS*, **50**, 209

Milne, D. K., & Webster, B. L. 1979, *A&AS*, **36**, 169

Minkowski, R. 1964, *PASP*, **76**, 197

Minkowski, R., & Aller, L. H. 1954, *ApJ*, **120**, 261

Minkowski, R., & Osterbrock, D. E. 1960, *ApJ*, **131**, 537

Monk, D. J., Barlow, M. J., & Clegg, R. E. S. 1988, *MNRAS*, **234**, 583

Moseley, H. 1980, *ApJ*, **238**, 892

Mufson, S. L., Lyon, J., & Marionni, P. A. 1975, *ApJ*, **201**, L85
Nagata, T., Tokunaga, A., Sellgren, K., Smith, R. G., Onaka, T., Nakada, Y., & Sakata, A. 1988, *ApJ*, **326**, 158
Nahar, S. N. 1995, *ApJS*, **101**, 423
Napiwotzki, R. 1993, *Acta Astron.*, **43**, 343
Napiwotzki, R. 1998, *Reviews in Modern Astronomy*, **11**, 3
Napiwotzki, R., & Schönberner, D. 1991, *A&A*, **249**, L16
Napiwotzki, R., & Schönberner, D. 1995, *A&A*, **301**, 545
Neri, R., Kahane, C., Lucas, R., Bujarrabal, V., & Loup, C. 1998, *A&A*, **130**, 1
Ney, E. P., Merrill, K. M., Becklin, E. E., Neugebauer, G., & Wynn-Williams, C. G. 1975, *ApJ*, **198**, L129
Nuth, J. A., Moseley, S. H., Silverberg, R. F., Goebel, J. H., & Moore, W. J. 1985, *ApJ*, **290**, L41
O'Dell, C. R. 1963, *ApJ*, **138**, 67
O'Dell, C. R., & Handron, K. D. 1996, *AJ*, **111**, 1630
Omont, A., Moseley, S. H., Forveille, T., Glaccum, W., Harvey, P. M., & Likkel, L. 1989, in *Proc. 22nd Eslab Symp, on Infrared Spectroscopy in Astronomy*, ESA SP-290, p. 379
Omont, A. *et al.* 1995, *ApJ*, **454**, 819
Oster, L. 1961, *ApJ*, **134**, 1010
Paczyński, B. 1970, *Acta Astr.* **20**, 47
Paczyński, B. 1971a, *Acta Astr.*, **21**, 417
Paczyński, B. 1971b, *Astrophys. Lett.*, **9**, 33
Papp, K. A., Purton, C. R., & Kwok, S. 1983, *ApJ*, **268**, 145
Parker, Q. E., & Phillipps, S. 1998, *Astronomy & Geophysics*, **39**, No.4, 10
Parthasarathy, M., & Pottasch, S. R. 1986, *A&A*, **154**, L16
Patriachi, P., & Perinotto, M. 1991, *A&AS*, **91**, 325
Pauldrach, A., Puls, J., Kudritzki, R. P., Mèndez, R. H., & Heap, S. R. 1988, *A&A*, **207**, 123
Peimbert, M. 1978, in *IAU Symp. 76: Planetary Nebulae*, ed. Y. Terzian, (Reidel:Dordrecht), p. 215
Percy, J. R., Sasselov, D. D., Alfred, A., & Scott, G. 1991 *ApJ*, **375**, 691
Perek, L., & Kohoutek, L. 1967, *Catalog of Galactic Planetary Nebulae*, (Academia:Praha)
Perinotto, M. 1991, *ApJS*, **76**, 687
Perinotto, M. 1993, in *IAU Symp 155: Planetary Nebulae*, eds. R. Weinberger & A. Acker (Kluwer:Dordrecht), p. 57
Perinotto, M., Kifonidis, K., Schönberner, D., & Marten, H. 1998, *A&A*, **332**, 1044
Perrine, C. D. 1929, *Astron. Nach.*, **237**, 89
Péquignot, D. 1997, in *IAU Symp 180: Planetary Nebulae*, eds. H. J. Habing & H. Lamers (Kluwer:Dordrecht), p. 167
Péquignot, D., & Baluteau, J.-P. 1994, *A&A*, **283**, 593
Plaskett, J. S. 1928, Harvard Circ. No. 335, 1928
Pottasch, S. R. 1980, *A&A*, **89**, 336
Pottasch, S. R. 1984, *Planetary Nebulae*, (Reidel:Dordrecht)
Pottasch, S. R. 1989, in *IAU Symp 131: Planetary Nebulae*, eds. S. Torres-Peimbert (Kluwer:Dordrecht), p. 481
Pottasch, S. R. 1990, *A&A*, **236**, 231
Pottasch, S. R. 1995, in *Asymmetric Planetary Nebulae, Ann. Israel Phys. Soc.*, **Vol. 11**, eds. A. Harpaz & N. Soker, p. 7
Pottasch, S. R. 1996, *Ap&SS*, **238**, 17
Pottasch, S. R. 1993, in *IAU Symp 155: Planetary Nebulae*, eds. R. Weinberger & A. Acker (Kluwer:Dordrecht), p. 449
Pottasch, S. R. 1997, in IAU Symp. 180: Planetary Nebulae, eds. H. J. Habing & H. Lamers (Kluwer:Dordrecht), p. 483
Pottasch, S. R., & Zijlstra, A. A. 1992, *A&A*, **256**, 251
Pottasch, S. R., & Zijlstra, A. A. 1994, *A&A*, **289**, 261
Pottasch, S. R., Goss, W. M., Gathier, R., & Arnal, E. M. 1982, *A&A*, **106**, 229
Pottasch, S. R. *et al.* 1984, *A&A*, **138**, 10
Preite-Martinez, A., & Pottasch, S. R. 1983, *A&A*, **126**, 31
Preite-Martinez, A., Acker, A., Köppen, J., & Stenholm, B. 1989, *A&AS*, **81**, 309
Preite-Martinez, A., Acker, A., Köppen, J., & Stenholm, B. 1991, *A&AS*, **88**, 121
Purton, C. R., Feldman, P. A., Marsh, K. A., Allen, D. A., & Wright, A. E. 1982, *MNRAS*, **198**, 321

Pwa, T. H., Pottasch, S. R., & Mo, J. E. 1986, *A&A*, **164**, 184

Racine, R., & Harris, W. E. 1989, *AJ*, **98**, 1609

Ramsbottom, C. A., Bell, K. L., & Stafford, R. P. 1996, *At. Data Nucl. Data Tables*, **63**, 57

Rana, N. C. 1991, *Ann. Rev. Astr. Ap.*, **29**, 129

Ratag, M. A. 1991, PhD. Thesis, University of Groningen

Reimers, D. 1975, *Mém. Soc. Roy. Sci. Liège*, 6e Ser. 8, p.369

Reimers, D., & Koester, D. 1982, *A&A*, **116**, 341

Renzini, A. 1982, in *IAU Symp 131: Planetary Nebulae*, eds. D. R. Flower, (Reidel:Dordrecht), p. 267

Renzini, A., & Buzzoni, A. 1986, in *Spectral Evolution of Galaxies*, eds. C. Chiosi, & Renzini (Reidel:Dordrecht), p. 195

Renzini, A., & Voli, M. 1981, *A&A*, **94**, 175

Richer, H. B. *et al.* 1997, *ApJ*, **484**, 741

Richer, M. G., & McCall, M. L. 1995, *ApJ*, **445**, 642

Richer, M. G., McCall, M. L., & Stasińska, G. 1998, *A&A*, **340**, 67

Rola, C., & Stasińska, G. 1994, *A&A*, **282**, 199

Romanishin, W., & Angel, J. R. P. 1980, *ApJ.*, **235**, 992

Roxburgh, I. W. 1967, *Nature*, **215**, 838

Russell, H. N., Dugan, R. M., & Stewart, J. Q. 1927, *Astronomy*, p. 837

Russell, R. W., Soifer, B. T., & Willner, S. P 1977, *ApJ*, **217**, L149

Russell, R. W., Soifer, B. T., & Willner, S. P. 1978, *ApJ*, **220**, 568

Sahai, R. *et al.* 1998a, *ApJ*, **493**, 301

Sahai, R. *et al.* 1998b, *ApJ*, **492**, L163

Salpeter, E. E. 1955, *ApJ*, **121**, 161

Salpeter, E. E. 1968, *IAU Symp. 34: Planetary Nebulae*, eds. D. Osterbrock & C. R. O'Dell (Reidel:Dordrecht), p. 409

Sasselov, D. D. 1984. *Ap&SS*, **102**, 161

Schmidt, G. D., & Cohen, M. 1981, *ApJ*, **246**, 444

Schmidt-Voigt, M. & Köppen, J. 1987a, *A&A*, **174**, 211

Schmidt-Voigt, M., & Köppen, J. 1987b, *A&A*, **174**, 223

Schneider, S. E., Silverglate, P. R., Altschuler, D. R., & Giovanardi, C. 1987, *ApJ*, **314**, 572

Schönberner, D. 1979, *A&A*, **79**, 108

Schönberner, D. 1981, *A&A*, **103**, 119

Schönberner, D. 1983, *ApJ*, **272**, 708

Schönberner, D. 1987, in *Late Stages of Stellar Evolution*, eds. S. Kwok & S. R. Pottasch (Reidel:Dordrecht), p. 337

Schönberner, D. 1993, *Acta Astron.*, **43**, 297

Schönberner, D. 1997, in *IAU Symp. 180: Planetary Nebulae*, ed. H. J. Habing & H. Lamers (Kluwer:Dordrecht), p. 379

Schönberner, D., & Napiwotzki, R. 1990, *A&A*, **231**, L33

Schönberner, D., & Weidemann, V. 1983, in *IAU Symposium 103: Planetary Nebulae*, ed. R. D. Flower (Reidel:Dordrecht), p. 359

Schönberner, D., Steffen, M., Stahlberg, J., Kifonidis, K., & Blöcker, T. 1997, in *Advances in Stellar Evolution*, eds. R. T. Rood & A. Renzini, Cambridge U. Press, p. 146

Schwarz, H. E., Corradi, R. L. M., & Melnick, J. 1992, *A&AS*, **96**, 23

Schwarzchild, & Härm 1967, *ApJ*, **150**, 961

Scott, P. F. 1973, *MNRAS*, **161**, 35P

Scott, P. F. 1975, *MNRAS*, **170**, 487

Seaton, M. J. 1954a, *MNRAS*, **114**, 54

Seaton, M. J. 1954b, *MNRAS* **118**, 154

Seaton, M. J. 1966, *MNRAS*, **132**, 113

Sharpless, S. 1959, *ApJS*, **4**, 257

Shaviv, G. 1978, *IAU Symp 76: Planetary Nebulae*, eds. Y. Terzian, (Reidel:Dordrecht), p. 195

Shaw, R. A., & Kaler, J. B. 1989, *ApJS*, **69**, 495

Shibata, K. M., Deguchi, S., Kasuga, T., Tamura, S., Nirano, N., & Kameya, O. 1994, in *IAU Colloq. 140: Astronomy with Millimeter and Submillimeter Wave Interferometry*, eds. M. Ishiguro & W. J. Welch, ASP Conf. Ser. Vol. **59**, p. 143

Shields, G. A. 1980, *PASP*, **92**, 418

Shklovsky, I. 1956a, *Astr. Zh.*, **33**, 222

Shklovsky, I. 1956b, *Astr. Zh.*, **33**, 315

Shortley, G., Aller, L. H., Baker, J. G., & Menzel, D. H. 1941, *ApJ*, **93**, 178

Skinner, C. J., Meixner, M. M., Hawkins, G. W., Keto, E., Jernigan, J. G., & Arens, J. F. 1994, *ApJ*, **423**, L135

Slijkhuis, S. 1992, PhD. thesis, Univ. Amsterdam, Netherlands

Smith, L. F., & Aller, L. H. 1969, *ApJ*, **157**, 1245

Smith, V. V., & Lambert, D. L. 1990, *ApJS*, **72**, 387

Smith, R. L., & Rose, W. K. 1972, *ApJ*, **176**, 395

Soker, N., & Livio, M. 1989, *ApJ*, **339**, 268

Solomon, P., Jefferts, K. B., Penzias, A. A., & Wilson, R. W. 1971, *ApJ*, **163**, L53

Sourisseau, C., Coddens, G., & Papoular, R. 1992, *A&A*, **254**, L1

Stanghellini, L., & Pasquali, A. 1995, *ApJ*, **452**, 286

Stanghellini, L., Corradi, R. L. M., & Schwarz, H. E. 1993, *A&A*, **279**, 521

Stasińska, G., & Tylenda, R. 1990, *A&A*, **240**, 467

Stasińska, G., Tylenda, R., Acker, A., & Stenholm, B. 1991, *A&A*, **247**, 173

Stephenson, C. B. 1989, *Publ. Warner & Swasey Obs.*, **3**, No. 2

Storey, P. 1981, *MNRAS*, **195**, 27P

Stoy, R. W. 1933, *MNRAS*, **93**, 588

Strömgren, B. 1939, *ApJ*, **89**, 526

Sun, J., & Kwok, S. 1987, *A&A*, **185**, 258

Szczerba, R. 1990, *A&A*, 237, 495

Tarafdar, S. P., & Apparao, K. M. V. 1988, *ApJ*, **327**, 342

Telesco, C. M., & Harper, D. A. 1977, *ApJ*, **211**, 475

te Lintel Hekkert, P., Caswell, J. L., Habing, H. J., Haynes, R. F., & Norris, R. P. 1991, *A&AS*, **90**, 327

te Lintel Hekkert, P., Chapman, J. M., & Zijlstra, A. A. 1992, *ApJ*, **390**, L23

Terzian, Y. 1966, *ApJ*, **144**, 657

Terzian, Y. 1997, in *IAU Symp. 180: Planetary Nebulae*, eds. H. J. Habing & H. Lamers (Kluwer:Dordrecht), p. 29

Terzian, Y., Balick, B., & Bignell, C. 1974, *ApJ*, **188**, 257

Thé, P. S. 1962, *Contr. Bosscha. Obs.*, **14**, table V

Thronson, H. A., Latter, W. B., Black, J. H., Bally, J., & Hacking, P. 1987, *ApJ*, **322**, 770

Trams, N. R. 1991. PhD. thesis. Rijksuniv. Utrecht, Netherlands

Trams, N. F., Waters, L. B. F. M., Waelkens, C., Lamers, H. J. G. L. M., & van der Veen, W. E. C. J. 1989, *A&A*, **218**, L1

Treffers, R., & Cohen, M. 1974, *ApJ*, **188**, 545

Tremblay, M., & Kwok, S. 1997 in *IAU Symp. 180: Planetary Nebulae*, eds. H. J. Habing & H. Lamers (Kluwer:Dordrecht), p. 413

Trimble, V., & Sackman, I.-J. 1978, *MNRAS*, **182**, 97

Tuchman, Y., & Barkat, Z. 1980, *ApJ*, **242**, 199

Tuchman, Y., Sack, N., & Barket, Z. 1979, *ApJ*, **234**, 217

Tylenda, R. 1983, *A&A*, **126**, 299

Tylenda, R. 1989, in *IAU Symp. 131: Planetary Nebulae*, eds. S. Torres-Peimbert (Kluwer:Dordrecht), p. 531

Tylenda, R., & Stasińska, G. 1994, *A&A*, **288**, 897

Tylenda, R., Acker, A., & Stenholm, B. 1993, *A&AS*, **102**, 595

Tylenda, R., Stansińska, G., Acker, A., & Stenholm, B. 1991, *A&A*, **246**, 221

Tylenda, R., Acker, A., Stenholm, B., & Köppen, J. 1992, *A&AS*, **95**, 337

van der Hucht, K. A., Jurriens, T. A., Olnon, F. M., Thé, P. S., Wesselius, P. R., & Williams, P. M. 1985, *A&A*, **145**, L13

van de Hulst, H. C. 1957 *Light Scattering by small particles*, (Dover: New York)

van de Steene, G. C., & Zijlstra, A. A. 1995, *A&A*, **293**, 541

van der Steene, G. C., Jacoby, G. H., & Pottasch, S. R. 1995, *A&AS*, **118**, 243

van der Steene, G. C., Sahu, K. C., & Pottasch, S. R. 1996, *A&AS*, **120**, 111

van der Veen, W. E. C. J., Habing, H. J., van Langevelde, H. J., & Geballe, T. R. 1989, *A&A*, **216**, L1

van der Veen, W. E. C. J., Habing, H. J., & Geballe, T. R. 1989, *A&A*, **226**, 108

van der Veen, W. E. C. J., Waters, R., Trams, N., & Matthews, H. E. 1994, *A&A*, **285**, 551 van Winckel, H., Waelkens, C., & Waters, L. B. F. M. 1995, *A&A*, **293**, L25

van Winckel, H. 1999, in *IAU Symp. 191: Asymptotic Giant Branch Stars*, eds. T. Le Bertre, A. Lébre, & C. Waelkens (San Francisco: ASP), 465

Vassiliadis, E. 1992, Thesis, Australian National University

Vassiliadis, E., & Wood, P. R. 1993, *ApJ*, **413**, 641

Vassiliadis, E., & Wood, P. R. 1994, *ApJS*, **92**, 125

Volk, K. 1992, *ApJS*, **80**, 347

Volk, K., & Kwok, S. 1985, *A&A*, **153**, 79

Volk, K., & Kwok, S. 1988, *ApJ*, **331**, 435

Volk, K., & Kwok, S. 1989, *Ap J*, **342**, 345

Volk, K., & Kwok, S. 1997, *ApJ*, **477**, 722

Volk, K., Kwok, S., & Hrivnak, B. J. 1999, *ApJ*, **516**, L99

Volk, K., Kwok, S., & Langill, P. P. 1992, *ApJ*, **391**, 285

Vorontsov-Velyaminov, B. A. 1934, *Astr. Zh.*, **11**, 40

Waelkens, C., Waters, L. B. F. M., Cassatella, A., Le Bertre, T., & Lamers, H. J. G. L. M. 1987, *A&A*, **181**, L5

Waelkens, C., van Winckel, H., Bogaert, E., & Trams, N. R. 1991, *A&A*, **251**, 495

Wannier, P. G., & Sahai, R. 1986, *ApJ*, **311**, 335

Wannier, P. G., & Sahai, R. 1987, *ApJ*, **319**, 367

Waters, L. B. F. M., Beintema, D. A., Zijlstra, A. A., de Koter, A., Molster, F. J., Bouwman, J., de Jong, T., Pottasch, S. R., & de Graauw, Th. 1998, *A&A*, **331**, L61

Webster A. 1995 *MNRAS* **277**, 1555

Webster, B. L. 1988, *MNRAS*, **230**, 377

Webster, B. L., & Glass, I. S. 1974, *MNRAS*, **166**, 491

Weidemann, V. 1987, *A&A*, **188**, 74

Weidemann, V. 1990, *Ann. Rev. Astr. Ap.*, **28**, 103

Weidemann V., & Koester, D. 1983, *A&A*, **121**, 77

Weidemann, V., Jordan, S., Iben, I., & Casertano, S. 1992, *AJ*, **104**, 1876

Weinberger, R. 1989, *A&AS*, **78**, 301

Wentzel, D. G. 1976 *ApJ*, **204**, 452

Werner, K., & Heber, U. 1991, in *Stellar Atmospheres: Beyond Classical Models*, eds. L. Crivellari *et al.* (Kluwer:London), p. 341

Wesemael, F., Green, R. F., & Liebert, J. 1985, *ApJS*, **58**, 379

Westbrook, W. E., Becklin, E. E., Merrill, K. M., Neugebauer, G., Schmidt, M. Willner, S. P., & Wynn-Williams, C. G. 1975. *ApJ*, **202**, 407

Westerlund, B., & Henize, K. G. 1967, *ApJS*, **14**, 154

Willson, L. A. 1981, in *Physical Processes in Red Giants*, eds. I. Iben & A. Renzini (Reidel:Dordrecht), p. 225

Willson, L. A. 1987, in *Late Stages of Stellar Evolution*, eds. S. Kwok & S. R. Pottasch (Reidel:Dordrecht), p. 253

Wilson, O. C. 1948, *ApJ*, **108**, 201

Wilson, O. C., & Aller, L. H. 1954, *ApJ*, **119**, 243

Wood, P. R. 1974, *ApJ*, **190**, 609

Wood, P. R. 1981, in *Physical Processes in Red Giants*, eds. I. Iben & A. Renzini (Reidel:Dordrecht), p. 205

Wood, M. 1992, *ApJ*, **386**, 539

Wood, P. R., & Faulkner, D. J. 1986, *ApJ*, **307**, 659

Wood, P. R., Bessell, M. S., & Dopita, M. A. 1986, *ApJ*, **311**, 632

Wood, P. R., Meatheringham, S. J., Dopita, M. A., & Morgan, D. H. 1987, *ApJ*, **320**, 178

Woolf, N. J., & Ney, N. 1969, *ApJ*, **155**, L181

Wright, W. H. 1918, *Lick Obs. Publ.*, **13**, 193

Yammaura, I., Onaka, T., Kamijo, F., Deguchi, S., & Ukita, N. 1996, *ApJ*, **465**, 926

Young, K., Serabyn, E., Phillips, T. G., Knapp, G. R., Güsten, R., & Schulz, A. 1992, *ApJ*, **385**, 265

Young, K., Keene, J., Phillips, T. G., Betz, A., & Boreiko, R. T. 1993, unpublished data

Young, K., Cox, P., Huggins, P. J., Forveille, T., & Bachiller, R. 1997, *ApJ*, **482**, L101

Zanstra, H. 1927, *ApJ*, **65**, 50

Zhang, C. Y. 1993, *ApJ*, **410**, 239

Zhang, C. Y. 1995, *ApJS*, **98**, 659

Zhang, C. Y., & Kwok, S. 1990, *A&A*, **237**, 479
Zhang, C. Y., & Kwok, S. 1991, *A&A*, **250**, 179
Zhang, C. Y., & Kwok, S. 1992, *ApJ*, **385**, 255
Zhang, C. Y., & Kwok, S. 1993, *ApJS*, **88**, 137
Zhang, C. Y., & Kwok, S. 1998, *ApJS*, **117**, 341
Zhang, C. Y., & Kwok, S. 1999, in preparation
Zhekov, S. A., & Perinotto, M. 1996, *A&A*, **309**, 648
Zijlstra, A. A., & Siebenmorgen, R. 1994, *Acta Astr.*, **43**, 409
Zijlstra, A. A., Pottasch, S. R., & Bignell, R. C. 1989, *A&AS*, **79**, 329
Zijlstra, A., te Lintel Hekkert, P., Pottasch, S. R., Caswell, J., Ratag, M., & Habing, H. 1989, *A&A*, **217**, 157

Appendix 1: List of symbols and abbreviations

α	the fine structure constant
α_A, α_B	total recombination coefficient under Cases A and B (cm^{-3} s^{-1})
α'	the rate of populating a state by direct recombination or cascade (cm^{-3} s^{-1})
α^{eff}	effective recombination coefficient (cm^{-3} s^{-1})
Γ	heating rate (erg cm^{-3} s^{-1})
ΔV	line width (cm s^{-1})
ϵ	filling factor
$\dot{\Theta}$	proper motion (rad s^{-1})
θ	angular radius
κ_ν	absorption coefficient (cm^{-1})
Λ	cooling rate (erg cm^{-3} s^{-1})
λ	wavelength (cm)
λ	time constant for star formation (s)
μ	electric dipole moment (esu cm)
μ	mean atomic weight per H atom
μ_e	mean atomic weight per electron
ν	frequency (Hz)
ρ_s	specific gravity of dust particle
ρ_0	space density of PN on the galactic plane (cm^{-3})
ρ_1	projected surface density of PN (cm^{-2})
σ	Stephan-Boltzmann constant (5.671 $\times$ 10^{-5} erg cm^{-2} s^{-1} K^{-4})
σ_n	recombination cross section (cm^2)
τ	optical depth
$\tau(m)$	main sequence lifetime of a star with mass m (s)
ϕ	thermal pulse phase
ϕ_ν	line profile (Hz^{-1})
χ	PN formation rate (cm^{-3} s^{-1})
ψ	the initial mass function (g^{-1})
Ω	solid angle (steradian)
A_{ij}	spontaneous emission coefficient (s^{-1})
a	radius of dust grain (cm)
a_0	the Bohr radius (0.529 $\times$ 10^{-8} cm^2)
a_ν	atomic absorption coefficient (cm^2 s^{-1} Hz^{-1})
B_ν	Planck's function (erg cm^{-2} s^{-1} ster^{-1} Hz^{-1})

B_{ij}, B_{ji}	stimulated emission and absorption coefficients (cm^2 erg^{-1} s^{-1})
C_{ij}	collisional rates (cm^{-3} s^{-1})
D	distance (cm)
E_H	energy released by converting 1 g of H to He (erg)
F_v	monochromatic flux (erg cm^{-2} s^{-1} Hz^{-1})
f	molecule to hydrogen abundance ratio
f_{12}	oscillator strength
$f(v)$	the Maxwellian distribution (s cm^{-1})
Gyr	10^9 yr
g	statistical weight
g	surface gravity (cm s^{-1})
$g(v, T)$	the Gaunt factor
H	hydrogen
h	Planck's constant (6.626 10^{-27} erg s)
HAC	hydrogenated amorphous carbon
He	helium
H-R diagram	Herztsprung-Russell diagram
I_v	specific intensity (erg cm^{-2} s^{-1} Hz^{-1} $ster^{-1}$)
i, j	upper and lower states of a b-b transition
IMF	initial mass function
IRE	infrared excess
J	rotational quantum number
J_v	average intensity (erg cm^{-2} s^{-1})
Jy	flux unit (erg cm^{-2} s^{-1} Hz^{-1})
j	integrated emission coefficient (erg cm^{-3} s^{-1} $ster^{-1}$)
j_v	emissivity (erg cm^{-3} s^{-1} $ster^{-1}$ Hz^{-1})
k	the Boltzmann constant (1.38×10^{-16} erg K^{-1})
k	inverse scale height (cm^{-1})
k	quantum number of a free state
L_*	luminosity of the central star (erg s^{-1})
$L_\odot$	solar luminosity (3.826×10^{33} erg s^{-1})
LRS	low resolution spectrometer
l	length element (cm)
ℓ	orbital angular momentum quantum number
M_c	core mass of AGB stars, mass of central stars of PN (g)
M_e	mass of the hydrogen envelope (g)
M_i	ionized mass of PN (g)
M_i	initial mass (g)
M_s	mass of the nebular shell (g)
$\dot{M}$	mass-loss rate of the AGB wind (g s^{-1})
$\dot{m}$	mass-loss rate of the central-star wind (g s^{-1})
$M_\odot$	solar mass (1.989×10^{33} g)
MS	main sequence
M_{WD}	upper MS mass limit for the formation of WD
n	principal quantum number

n_e	electron density in the nebula (cm^{-3})
n_p	proton density in the nebula (cm^{-3})
n_H	total H density in the nebula (cm^{-3})
P	period
PAH	polycyclic aromatic hydrocarbons
PN	planetary nebulae
pc	parsec $(3.086 \times 10^{18}$ cm$)$
$R(n, \ell)$	radial wavefunction with quantum numbers n and ℓ
R_i	radius of the ionized nebula (cm)
r_s	the Strømgren radius (cm)
S_ν	source function $(erg\ cm^{-2}\ s^{-1}\ Hz^{-1}\ ster^{-1})$
SED	spectral energy distribution
SFR	star formation rate
T_A	antenna temperature (K)
T_b	brightness temperature (K)
T_d	dust temperature (K)
T_e	kinetic temperature of electrons (K)
T_R	radiation tempreature (K)
T_x	excitation temperature (K)
V	velocity of the AGB wind $(cm\ s^{-1})$
V	volume (cm^3)
$V(b)$	visibility function $(erg\ cm^{-2}\ s^{-1}\ Hz^{-1})$
V_s	expansion velocity of the PN shell $(cm\ s^{-1})$
v	velocity of the central-star wind $(cm\ s^{-1})$
WD	white dwarf
x_e	electron to proton density ratio
Z	atomic number
Z	nuclear charge
Z	metallicity

Subject index